Solid Waste Management

Solid Waste Management

Editor: Karen Hardt

RCALLISTO REFERENCE

www.callistoreference.com

Callisto Reference,
118-35 Queens Blvd., Suite 400,
Forest Hills, NY 11375, USA

Visit us on the World Wide Web at:
www.callistoreference.com

ISBN: 978-1-64116-015-5 (Hardback)

Cataloging-in-Publication Data

Solid waste management / edited by Karen Hardt.
    p. cm.
Includes bibliographical references and index.
ISBN 978-1-64116-015-5
1. Refuse and refuse disposal. 2. Salvage (Waste, etc.).
3. Factory and trade waste. I. Hardt, Karen.
TD791 . S65 2018
363.728--dc23

Table of Contents

Permissions

Index

Preface

Solid waste amounts to the largest part of waste produced in the world. Therefore, its management and disposal is necessary. Solid waste management is a process through which waste present is collected, transported, treated and disposed. The two most widely used and popular ways to dispose waste are incineration and land filling. A large amount of the waste is also recycled. Some of ways of recycling the waste are pyrolysis, energy recovery, resource recovery and biological reprocessing. Most of the topics introduced in this book cover new techniques and the applications of solid waste management. This textbook will serve as a reference to a broad spectrum of readers.

A detailed account of the significant topics covered in this book is provided below:

Chapter 1- Due to a rise in consumption, there is an increase in waste. Developing a better understanding of the current situation would help in proper management of solid waste. One of the important ways of managing waste is to reduce materials that generate solid waste and to promote recycling. This chapter will provide an integrated understanding of solid waste management.

Chapter 2- This section focuses on the subject of waste generation and waste characterization. It is very important to determine the source, quantity and composition of waste. Solid waste management systems rely greatly on this data. The topics discussed in the chapter are of great importance to broaden the existing knowledge on waste generation and waste characterization.

Chapter 3- Solid waste collection does not merely mean collecting waste. Some of the aspects considered while collecting waste are public health and safety, collection route and storage containers. A transfer station is also necessary to act as a site of waste disposal. The aspects elucidated in this chapter are of vital importance, and provide a better understanding of solid waste collection.

Chapter 4- Waste that is collected needs to be properly disposed, as it can otherwise create health hazards and environmental issues. Some of the ways of waste disposal are sanitary landfill, composting, incineration, gasification, pyrolysis and refuse-derived fuel. This chapter is an overview of the subject matter incorporating all the major aspects of solid waste disposal.

Chapter 5- Processing waste requires proper techniques and equipments. It helps in recovering products and energy and in recovering materials for reuse. The process of recycling can help in decreasing consumption which would in return decrease the production of waste. This chapter has been carefully written to provide an easy understanding of the varied facets of solid waste treatment.

It gives me an immense pleasure to thank our entire team for their efforts. Finally in the end, I would like to thank my family and colleagues who have been a great source of inspiration and support.

Editor

A Brief Introduction to Solid Waste Management

Due to a rise in consumption, there is an increase in waste. Developing a better understanding of the current situation would help in proper management of solid waste. One of the important ways of managing waste is to reduce materials that generate solid waste and to promote recycling. This chapter will provide an integrated understanding of solid waste management.

Solid Waste Management (SWM)

Solid waste management (SWM) is associated with the control of waste generation, its storage, collection, transfer and transport, processing and disposal in amanner thatisin accordance with thebest principlesof publichealth, economics, engineering, conservation, aesthetics, public attitudeand other environmental considerations.

Put differently, the SWM processes differ depending on factors such as economic status (e.g., the ratio of wealth created by the production of primary products to that derived from manufactured goods, per capita income,etc.), degree of industrialisation, social development (e.g., education, literacy, healthcare, etc.) and quality of life of a location. In addition, regional, seasonal and economic differences influence the SWM processes. This, therefore, warrants management strategies that areeconomicallyviable,technically feasible and socially acceptable to carry out such of the functions as are listed below:

- Protection of environmental health.

- Promotion of environmental quality.

- Supporting the efficiency and productivity of the economy.

- Generation of employment and income.

SWM has socio-economic and environmental dimensions. In the socio-economic dimension, for example, it includes various phases such as waste storage, collection, transport and disposal, and the management of these phases has to be integrated. In other words, wastes have to be properly stored, collected and disposed of by co-operative management. In addition, poor management of wastes on the user side such as disposing of wastes in the streets, storm water drains, rivers and lakes has to be avoided to preserve the environment, control vector-born diseases and ensure water quality/resource.

Municipal Solid Waste

Municipal solid waste (MSW), commonly known as trash or garbage in the United States and as refuse or rubbish in Britain, is a waste type consisting of everyday items that are discarded by the public. "Garbage" can also refer specifically to food waste, as in a garbage disposal; the two are sometimes collected separately.

Composition

The composition of municipal solid waste varies greatly from municipality to municipality and changes significantly with time. In municipalities which have a well developed waste recycling system, the waste stream consists mainly of intractable wastes such as plastic film, and non-recyclable packaging materials. At the start of the 20th century, the majority of domestic waste (53%) in the UK consisted of coal ash from open fires. In developed areas without significant recycling activity it predominantly includes food wastes, market wastes, yard wastes, plastic containers and product packaging materials, and other miscellaneous solid wastes from residential, commercial, institutional, and industrial sources. Most definitions of municipal solid waste do not include industrial wastes, agricultural wastes, medical waste, radioactive waste or sewage sludge. Waste collection is performed by the municipality within a given area. The term *residual waste* relates to waste left from household sources containing materials that have not been separated out or sent for reprocessing.Waste can be classified in several ways but the following list represents a typical classification:

- Biodegradable waste: food and kitchen waste, green waste, paper (most can be recycled although some difficult to compost plant material may be excluded)

- Recyclable materials: paper, cardboard, glass, bottles, jars, tin cans, aluminum cans, aluminum foil, metals, certain plastics, fabrics, clothes, tires, batteries, etc.

- Inert waste: construction and demolition waste, dirt, rocks, debris

- Electrical and electronic waste (WEEE) - electrical appliances, light bulbs, washing machines, TVs, computers, screens, mobile phones, alarm clocks, watches, etc.

- Composite wastes: waste clothing, Tetra Packs, waste plastics such as toys

- Hazardous waste including most paints, chemicals, tires, batteries, light bulbs, electrical appliances, fluorescent lamps, aerosol spray cans, and fertilizers

- Toxic waste including pesticides, herbicides, and fungicides

- Biomedical waste, expired pharmaceutical drugs, etc.

Components of Solid Waste Management

The municipal solid waste industry has four components: recycling, composting, disposal, and waste-to-energy via incineration. There is no single approach that can be applied to the management of all waste streams, therefore the Environmental Protection Agency, federal agency of the United States of America, developed a hierarchy ranking strategy for municipal solid waste. The Waste Management Hierarchy is made up of four levels ordered from most preferred to least preferred methods based on their environmental soundness: Source reduction and reuse; recycling or composting; energy recovery; treatment and disposal.

Bins to collect paper, aluminium, glass, PET bottles and incinerable waste.

Collection

The functional element of collection includes not only the gathering of solid waste and recyclable materials, but also the transport of these materials, after collection, to the location where the collection vehicle is emptied. This location may be a materials processing facility, a transfer station or a landfill disposal site.

Waste Handling and Separation, Storage and Processing at the Source

Waste handling and separation involves activities associated with waste management until the waste is placed in storage containers for collection. Handling also encompasses the movement of loaded containers to the point of collection. Separating different types of waste components is an important step in the handling and storage of solid waste at the source.

Segregation and Processing and Transformation of Solid Wastes

The types of means and facilities that are now used for the recovery of waste materials that have been separated at the source include curbside ('kerbside' in the UK) collection, drop-off and buy-back centers. The separation and processing of wastes that have been separated at the source and the separation of commingled wastes usually occur at a materials recovery facility, transfer stations, combustion facilities and disposal sites.

Transfer and Transport

This element involves two main steps. First, the waste is transferred from a smaller collection vehicle to larger transport equipment. The waste is then transported, usually over long distances, to a processing or disposal site.

Disposal

Mixed municipal waste, Hiriya, Tel Aviv

Today, the disposal of wastes by land filling or land spreading is the ultimate fate of all solid wastes, whether they are residential wastes collected and transported directly to a landfill site, residual materials from materials recovery facilities (MRFs), residue from the combustion of solid waste, compost, or other substances from various solid waste processing facilities. A modern sanitary landfill is not a dump; it is an engineered facility used for disposing of solid wastes on land without creating nuisances or hazards to public health or safety, such as the problems of insects and the contamination of ground water.

Reusing

In the recent years environmental organizations, such as Freegle or Freecycle Network, have been gaining popularity for their online reuse networks. These networks provide a worldwide online registry of unwanted items that would otherwise be thrown away, for individuals and nonprofits to reuse or recycle. Therefore, this free Internet-based service reduces landfill pollution and promotes the gift economy.

Landfills

Landfills are created by land dumping. Land dumping methods vary, most commonly it involves the mass dumping of waste into a designated area, usually a hole or sidehill. After the waste is dumped, it is then compacted by large machines. When the dumping cell is full, it is then "sealed" with a plastic sheet and covered in several feet of dirt. This is the primary method of dumping in the United States because of the low cost and abundance of unused land in North America. Landfills pose the threat of pollution, and can intoxicate ground water. The signs of pollution are effectively masked by disposal companies and it is often hard to see any evidence. Usually landfills are surrounded by large walls or fences hiding the mounds of debris. Large amounts of chemical odor eliminating agent are sprayed in the air surrounding landfills to hide the evidence of the rotting waste inside the plant.

Energy Generation

Municipal solid waste can be used to generate energy. Several technologies have been developed that make the processing of MSW for energy generation cleaner and more economical than ever before, including landfill gas capture, combustion, pyrolysis, gasification, and plasma arc *gasification*. While older waste incineration plants emitted a lot of pollutants, recent regulatory changes and new technologies have significantly reduced this concern. United States Environmental Protection Agency (EPA) regulations in 1995 and 2000 under the Clean Air Act have succeeded in reducing emissions of dioxins from waste-to-energy facilities by more than 99 percent below 1990 levels, while mercury emissions have been reduced by over 90 percent. The EPA noted these improvements in 2003, citing waste-to-energy as a power source "with less environmental impact than almost any other source of electricity".

SWM System

A SWM system refers to a combination of various functional elements associated with the management of solid wastes. The system, when put in place, facilitates the collection and disposal of solid wastes in the community at minimal costs,while preserving public health and ensuring little or minimal adverse impact on the environment. The functional elements that constitute the system are:

(i) Waste generation: Wastes are generated at the start of any process, and thereafter, at every stage as raw materials are converted into goodsfor consumption. The source of waste generation, as we touched upon earlier, determines quantity, composition and waste characteristics. For example, wastes are generated from households, commercial areas, industries, institutions, street cleaning and other municipal services. The most important aspect of this part of the SWM systemisthe identification of waste.

(ii) Waste storage: Storage is a key functional element because collection of wastes never takes place at the source or at the time of their generation. The heterogeneous wastes generated in residential areas must be removed within 8 days due to shortage of storage space and presence of biodegradable material. Onsite storage is of primary importance due to aesthetic consideration, public health and economics involved. Some of the options for storage are plastic containers, conventional dustbins (of households), used oil drums, large storage bins (for institutions and commercial areas or servicing depots), etc. Obviously, these vary greatly in size, form and material.

(iii) Waste collection: This includes gathering of wastes and hauling them to the location, where the collection vehicle is emptied, which may be a transfer station (i.e., intermediate station where wastes from smaller vehicles are transferred to larger ones and also segregated), a processing plant or a disposal site. Collection depends on the number of containers, frequency of collection, types of collection services and routes. Typically, collection is provided under various management arrangements, ranging from municipal services to franchised services, and under various forms of contracts.

Note that the solution to the problem of hauling is complicated. For instance, vehicles used for long distance hauling may not be suitable or particularly economic for house-to-house collection. Every SWM system, therefore, requires an individual solution to its waste collection problem.

(iv) Transfer and transport: This functional element involves:

- the transfer of wastes from smaller collection vehicles, where necessary to overcome the problemof narrow accesslanes, to larger onesat transfer stations;

- the subsequent transport of the wastes, usually over long distances, to disposal sites.

The factors that contribute to the designing of a transfer station include the type of transfer operation, capacity, equipment, accessories and environmental requirements.

(v) Processing: Processing is required to alter the physical and chemical characteristics of wastes for energy and resource recovery and recycling. The important processing techniques include compaction, thermal volume reduction, manual separation of waste components, incineration and composting.

(vi) Recovery and recycling: This includes various techniques, equipment and facilities used to improve both the efficiency of disposal system and recovery of usable material and energy. Recovery involves the separation of valuable resources from the mixed solid wastes, delivered at transfer stations or processing plants. It also involves size reduction and density separation by air classifier, magnetic device for iron and screens for glass. The selection of any recovery process is a function of economics, i.e., costs of separation versus the recovered-material products. Certain recovered materials likeglass, plastics, paper, etc., can be recycled as they have economic value.

(vii) Waste disposal: Disposal is the ultimate fate of all solid wastes, be they residential-wastes,semi-solidwastesfrommunicipalandindustrialtre atmentplants, incinerator residues, composts or other substances that have no further use to the society. Thus, land use planning becomes a primary determinant in the selection, design and operation of landfill operations.A modernsanitary landfill isa method of disposing solid waste without creating a nuisance and hazard to public health. Generally, engineering principles are followed to confine the wastesto the smallest possible area, reduce themto the lowest particle volume by compaction at the site and cover them after each day's operation to reduce exposure to vermin. One of the most important functional elements of SWM, therefore, relates to the final use of the reclaimed land.

Typical SWM System: Functional Elements

You must, however, note that all the elements and linkages shown in figure need not necessarily be always present in a SWM system. Being generic in its form, this system is applicable to all regions, irrespective of their relative state of development (Tchobanoglous, et al., 1977) .

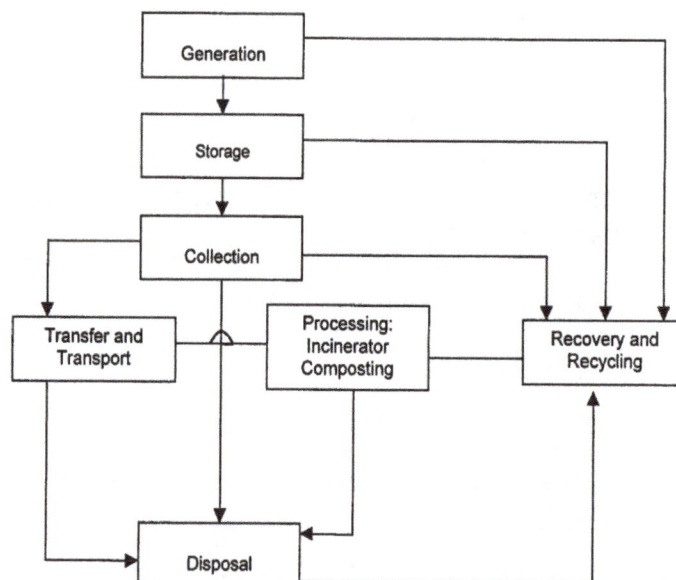

ESSWM and EST

We must recognise that each functional element discussed is closely interconnected to minimise adverse impact of wastes on the environment and to maximise the ecosystem carrying capacity. To derive optimal benefits from this, we must apply environmentally sound solid waste management (ESSWM) . This is an integrated approach for controlling and preserving the resources, both in quantity and quality. To improve environmental quality and achieve sustainable development, it is necessary we use EST – environmentally sound technologies (Matsumoto, et al., 2000).

Environmentally Sound Solid Waste management (ESSWM)

In any waste or resource management system, we must pay attention to the interaction between human activities and the ecosystem. We have to recognise that human activities including consumption of goods/services, production of wastes, etc., have a serious impact on the carrying capacity of the ecosystem. This in turn affects human health, as the environment deteriorates. The fundamental principles of ESSWM, which take into account economic and social issues along with environmental impact consideration, include the following:

- To ensure sustainable development of the ecosystem and human environment.
- To minimise the impact of human activities on the environment.
- To minimise the impact on the environment and maximise the ecosystem's carrying capacity.
- To ensure the implementation of ESSWM through environmentally sound technologies.

Environmentally Sound Technologies (Est)

EST refers to cost effective and energy efficient technologies, which generally perform better on the environment, as they do not pollute the ecosystem's vital components such as air, land or water and consider their use, recycling or recovery of wastes. EST can be categorised broadly as follows:

Hard EST: This includes equipment, machines and other infrastructure with their material accessories to handle waste products and monitor/measure the quality of air, water and soil.

Soft EST: This supports and complements hard technologies and include nature-based technologies and management tools. Nature-based technologies include processes and mechanisms nature uses withinaspecificecosystem (such as vermin composting) and its carrying capacity, while management tools include system and procedures, policy and regulatory frameworks, and environmental performance standards and guidelines.

Note that, as implied above, hard and soft technologies complement one another to achieve the goal.

EST is selected based on the following generic criteria, the indicators of which may vary depending on the regions in which they are implemented:

- Affordability: This means low investment, reasonableness, maintenance- free and durability.
- Validity: This refers to effectiveness, easy operation and maintenance.

Sustainability: This means low impact, energy saving and cultural acceptability.

Examples of EST for Collection and Transfer of Waste

Set-out container is one of the major factors that most collection system depends on. This is usually a paper or plastic bag, or a metal or plastic garbage or kraft paper bags in a metal or wooden frame. Set-out containers of rural areas include bags, pots, plastic or paper bags, cane or reed baskets, concrete or brick vats, urns, boxes, clay jars, or any kind of container available.

Non-compactortrucksaremoreefficientandcost-effectivethancompactor trucks in small cities and in areas where wastes tend to be very dense and havelittle potential for compaction. The use of lighter, more energy- efficientbox- trucks, vans, and dump trucks can be appropriate for sparsely populated areas, where the main constraint on collection efficiency is distance.

Transfer trailers or compacting vehicles can carry larger volumes of MSW than regular collection trucks, which allow them to travel longer distances carrying more waste. This lowers fuel costs, increases labour productivity, and saves on vehicle wear.

Factors Affecting SWM System

Many factors influence the decision-making process in the implementation of a SWM system (Phelps, et al., 1995) . Some of the factors that need to be considered in developing a SWM system are listed below:

(i) Quantities and characteristics of wastes: The quantities of wastes generated generally depend on the income level of a family, as higher income category tends to generate larger quantity of wastes, compared to low-income category. The quantity ranges from about 0.25 to about 2.3 kg per person per day, indicating a strong correlation between waste production and per capita income. One of the measures of waste composition (and characteristics) is density, which rangesfrom150 kg/m3 to 600 kg/m3. Proportion of paper and packaging materials in the waste largely account for the differences. When this proportion is high, the density is low and vice versa. The wastes of high density reflect a relativelyhigh proportion of organic matter andmoistureandlower levels of recycling.

(ii) Climate and seasonal variations: There are regions in extreme north (> 70N Latitude) and south (> 60 S Latitude), where temperatures are very lowfor much of the year. In cold climates, drifting snow and frozen ground interfere with landfill operations, and therefore, trenches must be duginsummerand cover material stockpiled for winter use. Tropical climates, on the other hand, are subject to sharp seasonal variations from wet to dry season, which cause significant changes in the moisture content of solid waste, varying from less than50% in dry season to greater than 65% in wet months. Collection and disposal of wastes in the wet months are often problematic.

High temperatures and humidity cause solid wastes to decompose far more rapidly than they do in colder climates. The frequency of waste collection in high temperature and humid climates should, therefore, be higher than that in cold climates. In sub-tropical or desert climate, there is no significant variation in moisture content of wastes (due to low rainfall) and low production of leachate from sanitary landfill. High winds and wind blown sand and dust, however, cause special problems at landfill sites. While temperature inversions can cause airborne pollutants to be trapped near ground level, landfill sites can affect groundwater by altering the thermal properties of the soil.

(iii) Physical characteristics of an urban area: In urban areas (i.e., towns and cities), where the layout of streets and houses is such thataccessby vehicles is possible and door-to-door collection of solid wastes is the accepted norm either by large compaction vehicle or smaller vehicle. The picture is, however, quite different in the inner and older city areaswhere narrow lanes make service by vehiclesdifficult and often impossible. Added to this is the problem of urban sprawl in the outskirts (of the cities) wherepopulationis growing at an alarming rate. Access ways are narrow, unpaved and tortuous, and therefore, not accessible to collection vehicles. Problems of solid waste storage and collection are most acute in such areas.

(iv) Financial and foreign exchange constraints: Solid waste management accounts for sizeable proportions of the budgets of municipal corporations. This is allocated for capital resources, which go towards the purchase of equipments, vehicles, and fuel and labour costs. Typically, 10% to 40% of the revenues of municipalities are allocated to solid waste management. In regions where wage rates are low, the aim is to optimise vehicle productivity. Theunfavourable financial situation of some countries hinders purchase of equipment and vehicles, and this situation is further worsened by the acute shortage of foreign exchange. This means that the balance between the degree of mechanisation and the sizeof the labour force becomes a critical issue in arriving at the most cost-effective solution.

(v) Cultural constraints: In some regions, long-standing traditions preclude the intrusion of waste collection on the precincts of households, and therefore, influence the collection system. In others, where the tradition of caste persists, recruits to the labour force for street cleaning and handling of waste must be drawn from certain sections of the population, while others will not consent to placing storage bins in their immediate vicinity. Social norms of a community more often than not over-ride what many may consider rational solutions. Waste management should, therefore, be sensitive to such local patterns of living and consider these factors in planning, design and operation.

(vi) Management and technical resources: Solid waste management, to be successful, requires a wide spectrum of workforce in keeping with the demands of the system. The best system for a region is one which makes full use of indigenous crafts and professional skills and/or ensures that training programmes are in place to provide a self-sustaining supply of trained workforce.

Waste Management

Waste management or Waste disposal is all the activities and actions required to manage waste from its inception to its final disposal. This includes amongst other things, collection, transport, treatment and disposal of waste together with monitoring and regulation. It also encompasses the legal and regulatory framework that relates to waste management encompassing guidance on recycling etc.

The term normally relates to all kinds of waste, whether generated during the extraction of raw materials, the processing of raw materials into intermediate and final products, the consumption of final products, or other human activities, including municipal (residential, institutional, commercial), agricultural, and social (health care, household hazardous waste, sewage sludge) . Waste management is intended to reduce adverse effects of waste on health, the environment or aesthetics.

Waste management practices are not uniform among countries (developed and developing nations) ; regions (urban and rural area), and sectors (residential and industrial) .

Central Principles of Waste Management

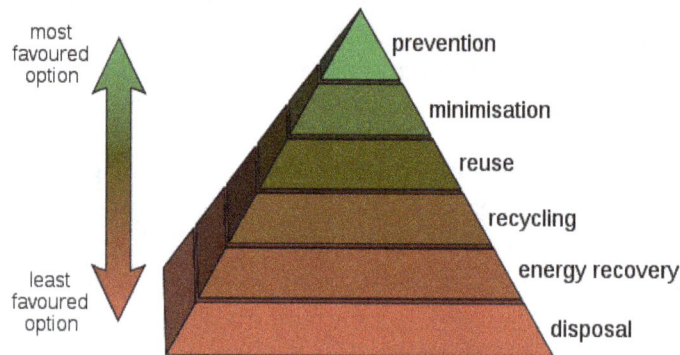

Diagram of the waste hierarchy

There are a number of concepts about waste management which vary in their usage between countries or regions. Some of the most general, widely used concepts include:

Waste Hierarchy

The waste hierarchy refers to the "3 Rs" reduce, reuse and recycle, which classify waste management strategies according to their desirability in terms of waste minimisation. The waste hierarchy remains the cornerstone of most waste minimisation strategies. The aim of the waste hierarchy is to extract the maximum practical benefits from products and to generate the minimum amount of waste. The waste hierarchy is represented as a pyramid because the basic premise is for policy to take action first and prevent the generation of waste. The next step or preferred action is to reduce the generation of waste i.e. by re-use. The next is recycling which would include composting. Following this step is material recovery and waste-to-energy. Energy can be recovered from processes i.e. landfill and combustion, at this level of the hierarchy. The final action is disposal, in landfills or through incineration without energy recovery. This last step is the final resort for waste which has not been prevented, diverted or recovered. The waste hierarchy represents the progression of a product or material through the sequential stages of the pyramid of waste management. The hierarchy represents the latter parts of the life-cycle for each product.

Life-cycle of a Product

The life-cycle begins with design, then proceeds through manufacture, distribution, use and then follows through the waste hierarchy's stages of reduce, reuse and recycle. Each of the above stages of the life-cycle offers opportunities for policy intervention, to rethink the need for the product, to redesign to minimize waste potential, to extend its use. The key behind the life-cycle of a product is to optimize the use of the world's limited resources by avoiding the unnecessary generation of waste.

Resource Efficiency

Resource efficiency reflects the understanding that current, global, economic growth and development can not be sustained with the current production and consumption patterns. Globally, we are extracting more resources to produce goods than the planet can replenish. Resource efficiency is the reduction of the environmental impact from the production and consumption of these goods, from final raw material extraction to last use and disposal. This process of resource efficiency can address sustainability.

Polluter Pays Principle

The Polluter pays principle is a principle where the polluting party pays for the impact caused to the environment. With respect to waste management, this generally refers to the requirement for a waste generator to pay for appropriate disposal of the unrecoverable material.

History

Throughout most of history, the amount of waste generated by humans was insignificant due to low population density and low societal levels of the exploitation of natural resources. Common waste produced during pre-modern times was mainly ashes and human biodegradable waste, and these were released back into the ground locally, with minimum environmental impact. Tools made out of wood or metal were generally reused or passed down through the generations.

However, some civilizations do seem to have been more profligate in their waste output than others. In particular, the Maya of Central America had a fixed monthly ritual, in which the people of the village would gather together and burn their rubbish in large dumps.

Modern Era

Sir Edwin Chadwick's 1842 report *The Sanitary Condition of the Labouring Population* was influential in securing the passage of the first legislation aimed at waste clearance and disposal.

Following the onset of industrialisation and the sustained urban growth of large population centres in England, the buildup of waste in the cities caused a rapid deterioration in levels of sanitation and the general quality of urban life. The streets became choked with filth due to the lack of waste clearance regulations. Calls for the establishment of a municipal authority with waste removal powers occurred as early as 1751, when Corbyn Morris in London proposed that "... as the preservation of the health of the people is of great importance, it is proposed that the cleaning of this city, should be put under one uniform public management, and all the filth be...conveyed by the Thames to proper distance in the country".

However, it was not until the mid-19th century, spurred by increasingly devastating cholera outbreaks and the emergence of a public health debate that the first legislation on the issue emerged. Highly influential in this new focus was the report *The Sanitary Condition of the Labouring Population* in 1842 of the social reformer, Edwin Chadwick, in which he argued for the importance of adequate waste removal and management facilities to improve the health and wellbeing of the city's population.

In the UK, the Nuisance Removal and Disease Prevention Act of 1846 began what was to be a steadily evolving process of the provision of regulated waste management in London. The Metropolitan Board of Works was the first citywide authority that centralized sanitation regulation for the rapidly expanding city and the Public Health Act 1875 made it compulsory for every household to deposit their weekly waste in "moveable receptacles: for disposal—the first concept for a dust-bin.

Manlove, Alliott & Co. Ltd. 1894 destructor furnace. The use of incinerators for waste disposal became popular in the late 19th century.

The dramatic increase in waste for disposal led to the creation of the first incineration plants, or, as they were then called, "destructors". In 1874, the first incinerator was built in Nottingham by Manlove, Alliott & Co. Ltd. to the design of Albert Fryer. However, these were met with opposition on account of the large amounts of ash they produced and which wafted over the neighbouring areas.

Similar municipal systems of waste disposal sprung up at the turn of the 20th century in other large cities of Europe and North America. In 1895, New York City became the first U.S. city with public-sector garbage management.

Early garbage removal trucks were simply open bodied dump trucks pulled by a team of horses. They became motorized in the early part of the 20th century and the first close body trucks to eliminate odours with a dumping lever mechanism were introduced in the 1920s in Britain. These were soon equipped with 'hopper mechanisms' where the scooper was loaded at floor level and then hoisted mechanically to deposit the waste in the truck. The Garwood Load Packer was the first truck in 1938, to incorporate a hydraulic compactor.

Waste Handling and Transport

Molded plastic, wheeled waste bin in Berkshire, England

Waste collection methods vary widely among different countries and regions. Domestic waste collection services are often provided by local government authorities, or by private companies for industrial and commercial waste. Some areas, especially those in less developed countries, do not have formal waste-collection systems.

Waste Handling Practices

Curbside collection is the most common method of disposal in most European countries, Canada, New Zealand and many other parts of the developed world in which waste is collected at regular intervals by specialised trucks. This is often associated with curb-side waste segregation. In rural areas waste may need to be taken to a transfer station. Waste collected is then transported to an appropriate disposal facility. In some areas, vacuum collection is used in which waste is transported from the home or commercial premises by vacuum along small bore tubes. Systems are in use in Europe and North America.

Pyrolysis is used for disposal of some wastes including tires, a process that can produce recovered fuels, steel and heat. In some cases tires can provide the feedstock for cement manufacture. Such systems are used in USA, California, Australia, Greece, Mexico, the United Kingdom and in Israel. The RESEM pyrolysis plant that has been operational at Texas USA since December 2011, and processes up to 60 tons per day. In some jurisdictions unsegregated waste is collected at the curb-side or from waste transfer stations and then sorted into recyclables and unusable waste. Such systems are capable of sorting large volumes of solid waste, salvaging recyclables, and turning the rest into bio-gas and soil conditioner. In San Francisco, the local government established its Mandatory Recycling and Composting Ordinance in support of its goal of zero waste by 2020, requiring everyone in the city to keep recyclables and compostables out of the landfill. The three streams are collected with the curbside "Fantastic 3" bin system – blue for recyclables, green for compostables, and black for landfill-bound materials – provided to residents and businesses and serviced by San Francisco's sole refuse hauler, Recology. The City's "Pay-As-You-Throw" system charges customers by the volume of landfill-bound materials, which provides a financial incentive to separate recyclables and compostables from other discards. The City's Department of the Environment's Zero Waste Program has led the City to achieve 80% diversion, the highest diversion rate in North

America. Other businesses such as Waste Industries use a variety of colors to distinguish between trash and recycling cans.

Financial Models

In most developed countries, domestic waste disposal is funded from a national or local tax which may be related to income, or notional house value. Commercial and industrial waste disposal is typically charged for as a commercial service, often as an integrated charge which includes disposal costs. This practice may encourage disposal contractors to opt for the cheapest disposal option such as landfill rather than the environmentally best solution such as re-use and recycling. In some areas such as Taipei, the city government charges its households and industries for the volume of rubbish they produce. Waste will only be collected by the city council if waste is disposed in government issued rubbish bags. This policy has successfully reduced the amount of waste the city produces and increased the recycling rate.

Disposal Solutions

Landfill

A landfill compaction vehicle in action.

Spittelau incineration plant in Vienna

Incineration

Incineration is a disposal method in which solid organic wastes are subjected to combustion so as to convert them into residue and gaseous products. This method is useful for disposal of residue of both solid waste management and solid residue from waste water management. This process reduces the volumes of solid waste to 20 to 30 percent of the original volume. Incineration and other high temperature waste treatment systems are sometimes described as "thermal treatment". Incinerators convert waste materials into heat, gas, steam, and ash.

Incineration is carried out both on a small scale by individuals and on a large scale by industry. It is used to dispose of solid, liquid and gaseous waste. It is recognized as a practical method of disposing of certain hazardous waste materials (such as biological medical waste) . Incineration is a controversial method of waste disposal, due to issues such as emission of gaseous pollutants.

Incineration is common in countries such as Japan where land is more scarce, as these facilities generally do not require as much area as landfills. Waste-to-energy (WtE) or energy-from-waste (EfW) are broad terms for facilities that burn waste in a furnace or boiler to generate heat, steam or electricity. Combustion in an incinerator is not always perfect and there have been concerns about pollutants in gaseous emissions from incinerator stacks. Particular concern has focused on some very persistent organic compounds such as dioxins, furans, and PAHs, which may be created and which may have serious environmental consequences.

Recycling

Waste not the Waste. Sign in Tamil Nadu, India

Steel crushed and baled for recycling

Recycling is a resource recovery practice that refers to the collection and reuse of waste materials such as empty beverage containers. The materials from which the items are made can be reprocessed into new products. Material for recycling may be collected separately from general waste using dedicated bins and collection vehicles, a procedure called kerbside collection. In some communities, the owner of the waste is required to separate the materials into different bins (e.g. for paper, plastics, metals) prior to its collection. In other communities, all recyclable materials are placed in a single bin for collection, and the sorting is handled later at a central facility. The latter method is known as "single-stream recycling."

The most common consumer products recycled include aluminium such as beverages cans, copper such as wire, steel from food and aerosol cans, old steel furnishings or equipment, rubber tyres, polyethylene and PET bottles, glass bottles and jars, paperboardcartons, newspapers, magazines and light paper, and corrugated fiberboard boxes.

PVC, LDPE, PP, and PS are also recyclable. These items are usually composed of a single type of material, making them relatively easy to recycle into new products. The recycling of complex products (such as computers and electronic equipment) is more difficult, due to the additional dismantling and separation required.

The type of material accepted for recycling varies by city and country. Each city and country has different recycling programs in place that can handle the various types of recyclable materials. However, certain variation in acceptance is reflected in the resale value of the material once it is reprocessed.

Re-use

Biological Reprocessing

An active compost heap.

Recoverable materials that are organic in nature, such as plant material, food scraps, and paper products, can be recovered through composting and digestion processes to decompose the organic matter. The resulting organic material is then recycled as mulch or compost for agricultural or landscaping purposes. In addition, waste gas from the process (such as methane) can be captured and used for generating electricity and heat (CHP/cogeneration) maximising efficiencies. The intention of biological processing in waste management is to control and accelerate the natural process of decomposition of organic matter.

Energy Recovery

Energy recovery from waste is the conversion of non-recyclable waste materials into usable heat, electricity, or fuel through a variety of processes, including combustion, gasification, pyrolyzation, anaerobic digestion, and landfill gas recovery. This process is often called waste-to-energy. Energy recovery from waste is part of the non-hazardous waste management hierarchy. Using energy recovery to convert non-recyclable waste materials into electricity and heat, generates a renewable energy source and can reduce carbon emissions by offsetting the need for energy from fossil sources as well as reduce methane generation from landfills. Globally, waste-to-energy accounts for 16% of waste management.

The energy content of waste products can be harnessed directly by using them as a direct combustion fuel, or indirectly by processing them into another type of fuel. Thermal treatment ranges from using waste as a fuel source for cooking or heating and the use of the gas fuel, to fuel for boilers to generate steam and electricity in a turbine. Pyrolysis and gasification are two related forms of thermal treatment where waste materials are heated to high temperatures with limited oxygen availability. The process usually occurs in a sealed vessel under high pressure. Pyrolysis of solid waste converts the material into solid, liquid and gas products. The liquid and gas can be burnt to produce energy or refined into other chemical products (chemical refinery) . The solid residue (char) can be further refined into products such as activated carbon. Gasification and advanced Plasma arc gasification are used to convert organic materials directly into a synthetic gas (syngas) composed of carbon monoxide and hydrogen. The gas is then burnt to produce electricity and steam. An alternative to pyrolysis is high temperature and pressure supercritical water decomposition (hydrothermal monophasic oxidation) .

Pyrolysis

Pyrolysis is a process of thermo-chemical decomposition of organic materials by heat in the absence of oxygen which produces various hydrocarbon gases. During pyrolysis, the molecules of object are subjected to very high temperatures leading to very high vibrations. Therefore, every molecule in the object is stretched and shaken to an extent that molecules starts breaking down. The rate of pyrolysis increases with temperature. In industrial applications, temperatures are above 430 °C (800 °F) . Fast pyrolysis produces liquid fuel for feedstocks like wood. Slow pyrolysis produces gases and solid charcoal. Pyrolysis hold promise for conversion of waste biomass into useful liquid fuel. Pyrolysis of waste plastics can produce millions of litres of fuel. Solid products of this process contain metals, glass, sand and pyrolysis coke which cannot be converted to gas in the process.

Resource Recovery

Resource recovery is the systematic diversion of waste, which was intended for disposal, for a specific next use. It is the processing of recyclables to extract or recover materials and resources, or convert to energy. These activities are performed at a resource recovery facility. Resource recovery is not only environmentally important, but it is also cost effective. It decreases the amount of waste for disposal, saves space in landfills, and conserves natural resources.

Resource recovery (as opposed to waste management) uses LCA (life cycle analysis) attempts to offer alternatives to waste management. For mixed MSW (Municipal Solid Waste) a number of broad studies have indicated that administration, source separation and collection followed by reuse and recycling of the non-organic fraction and energy and compost/fertilizer production of the organic material via anaerobic digestion to be the favoured path.

As an example of how resource recycling can be beneficial, many of the items thrown away contain precious metals which can be recycled to create a profit, such as the components in circuit boards. Other industries can also benefit from resource recycling with the wood chippings in pallets and other packaging materials being passed onto sectors such as the horticultural profession. In this instance, workers can use the recycled chips to create paths, walkways, or arena surfaces.

Sustainability

The management of waste is a key component in a business' ability to maintaining ISO14001 accreditation. Companies are encouraged to improve their environmental efficiencies each year by eliminating waste through resource recovery practices, which are sustainability-related activities. One way to do this is by shifting away from waste management to resource recovery practices like recycling materials such as glass, food scraps, paper and cardboard, plastic bottles and metal. This topic was on the agenda of the international Conference on Green Urbanism, held in Italy 12–14 October 2016.

Avoidance and Reduction Methods

An important method of waste management is the prevention of waste material being created, also known as waste reduction. Methods of avoidance include reuse of second-hand products, repairing broken items instead of buying new, designing products to be refillable or reusable (such as cotton instead of plastic shopping bags), encouraging consumers to avoid using disposable products (such as disposable cutlery), removing any food/liquid remains from cans and packaging, and designing products that use less material to achieve the same purpose (for example, lightweighting of beverage cans).

International Waste Movement

While waste transport within a given country falls under national regulations, trans-boundary movement of waste is often subject to international treaties. A major concern to many countries in the world has been hazardous waste. The Basel Convention, ratified by 172 countries, deprecates movement of hazardous waste from developed to less developed countries. The provisions of the Basel convention have been integrated into the EU waste shipment regulation. Nuclear waste, although considered hazardous, does not fall under the jurisdiction of the Basel Convention.

Benefits

Waste is not something that should be discarded or disposed of with no regard for future use. It can be a valuable resource if addressed correctly, through policy and practice. With rational and consistent waste management practices there is an opportunity to reap a range of benefits. Those benefits include:

1. Economic – Improving economic efficiency through the means of resource use, treatment and disposal and creating markets for recycles can lead to efficient practices in the production and consumption of products and materials resulting in valuable materials being recovered for reuse and the potential for new jobs and new business opportunities.

2. Social – By reducing adverse impacts on health by proper waste management practices, the resulting consequences are more appealing settlements. Better social advantages can lead to new sources of employment and potentially lifting communities out of poverty especially in some of the developing poorer countries and cities.

3. Environmental – Reducing or eliminating adverse impacts on the environmental through reducing, reusing and recycling, and minimizing resource extraction can provide improved air and water quality and help in the reduction of greenhouse gas emissions.

4. Inter-generational Equity – Following effective waste management practices can provide subsequent generations a more robust economy, a fairer and more inclusive society and a cleaner environment.

Challenges in Developing Countries

Waste management in cities with developing economies and economies in transition experience exhausted waste collection services, inadequately managed and uncontrolled dumpsites and the problems are worsening. Problems with governance also complicate the situation. Waste management, in these countries and cities, is an ongoing challenge and many struggle due to weak institutions, chronic under-resourcing and rapid urbanization. All of these challenges along with the lack of understanding of different factors that contribute to the hierarchy of waste management, affect the treatment of waste.

Technologies

Traditionally the waste management industry has been a late adopter of new technologies such as RFID (Radio Frequency Identification) tags, GPS and integrated software packages which enable better quality data to be collected without the use of estimation or manual data entry.

References

- Claire Swedberg (4 February 2014). "Air-Trak Brings Visibility to Waste Management". RFID Journal. Retrieved 1 October 2015

- Guidelines for National Waste Management Strategies Moving from Challenges to Opportunities (PDF). United Nations Environmental Programme. 2013. ISBN 978-92-807-3333-4

- Davidson, Gary (June 2011). "Waste Management Practices: Literature Review" (PDF). Dalhousie University - Office of Sustainability. Retrieved 3 March 2017

- "Editorial Board/Aims & Scope". Waste Management. 34 (3): IFC. March 2014. doi:10.1016/S0956-053X(14)00026-9

- Albert, Raleigh (4 August 2011). "The Proper Care and Use of a Garbage Disposal". Disposal Mag. Retrieved 2017-03-03

Waste Generation and Waste Characterization

This section focuses on the subject of waste generation and waste characterization. It is very important to determine the source, quantity and composition of waste. Solid waste management systems rely greatly on this data. The topics discussed in the chapter are of great importance to broaden the existing knowledge on waste generation and waste characterization.

Waste Stream Assessment (WSA)

Waste stream assessment (WSA) is a means to determine the basic aspects of quantity (i.e., the amount of waste generated in the community, both in terms of weight and volume), composition (i.e., the different components of waste stream) and sources of wastes.The information relating to these basic aspects of wastes is vital for making decisions about the SWM system, finance and regulations. Put differently, an assessment of waste stream is essential in the analyses of short- and long-term problems within the local waste management system. It also helps in targeting waste management activities and setting goals for different elements of a waste management plan.

Waste stream assessment, however, is not a one-time activity. It is a continuous and dynamic process, because the characteristics of wastes differ depending on the regions, communities, seasons, etc.

Rationale for Analysis

The reasons for the analysis of waste composition, characteristics and quantity include the following (Phelps, et al., 1995) :

(i) It provides the basic data for the planning, designing and operation of the management systems.

(ii) An ongoing analysis of the data helps detect changes in composition, characteristics and quantities of wastes, and the rates at which these changes take place, which facilitates effective implementation of management systems.

(iii) It quantifies the amount and type of materials suitable for processing, recovery and recycling.

(iv) It provides information that helps in deciding appropriate technologies and equipment.

(v) The forecast trends assist designers and manufacturers in the production of collection vehicles and equipment suitable for future needs.

In the absence of a reliable basic data, carrying out field investigations becomes necessary (Phelps et al., 1995).

Field Investigation

Field investigations may take any one or a combination of the following forms:

(i) Waste sorting: Sorting of wastes into predetermined components takes place at disposal sites for weighing and sampling in order to determine the percentage of each component and the physical and chemical characteristics of wastes. It is carried out manually, and the sample size for analysisisbetween100and150kg.Theimplementsrequiredforthepurposeincludesorting-table,measuringbox,binsorboxestocontain sorted materials and platform weighing machine.

(ii) Vehicle weighing: Vehicles are weighed when they enter the disposal sites loaded, and exit the sites empty. The vehicle's front wheels are weighed first, followed by the rear wheels and the sum of the two gives the total weight. Weighing is carried out each day of the weighing period in order to determine the average weight. The weighing of loaded andunloaded vehicles is accomplished with a weighing scale or weighbridge. Ideally, the weighing scale should be operated during the entire period of operation of the disposal site, round the clock, if necessary. An electronic or a mechanical portable axle scale, with a capacity of 20 tonnes is suitable for the purpose. An electronic scale comprises two load-cell platforms and an electronic control and a display unit. The quantity of waste measured at disposal sites reflects a disposal factor ratherthanagenerationfactor, since the measurements do not include wastes that are:

- salvaged at the generation and disposal sites;

- disposed of in unauthorised places such as vacant plots, alleys, ditches, etc.;

- salvaged by collectors;

- lost during transport.

(iii) Field visits: This means visiting institutional and industrial sites to identify wastes being generated and disposal methods. Field visits involve visiting the facility, i.e., industry, institutions, etc., viewing the waste handling system and completing a questionnaire with the assistance of the plant manager or senior technical personnel who usually investigate wastes from industries and institutions. Collection of samples in sealed polythene bags follows for laboratory analysis to identify physical and chemical characteristics. Each sample may be in the range of 1.5 to 5 kg.

An assessment of waste stream, in essence, helps us identify components that require improvement for effective implementation of waste management programmes (EPA, 1989 and 1995).

Waste Generation and Composition

Information on waste quantity and composition is important in evaluating alternatives in terms of equipment, systems, plans and management programmes. For example, if wastes generated at a commercial facility consist of only paper products, the appropriate equipment are shredders and balers. Similarly, on the basis of quantity generated, we can plan appropriate means for separation, collection and recycling programmes. That is to say, the success of SWM depends on the appropriate assessment of quantity of wastes generated.

Waste Generation

Waste generation encompasses those activities in which waste, be it solid or semi-solid material, no longer has sufficient economic value for its possessor to retain it.

The processing of raw materials is the first stage when wastes are generated, and waste generation continues thereafter at every step in the process as raw materials are converted into final products for consumption. Figure below shows a simplified material-flow diagram indicating the path of generation of solid wastes (Tchobanoglous, et al., 1977) :

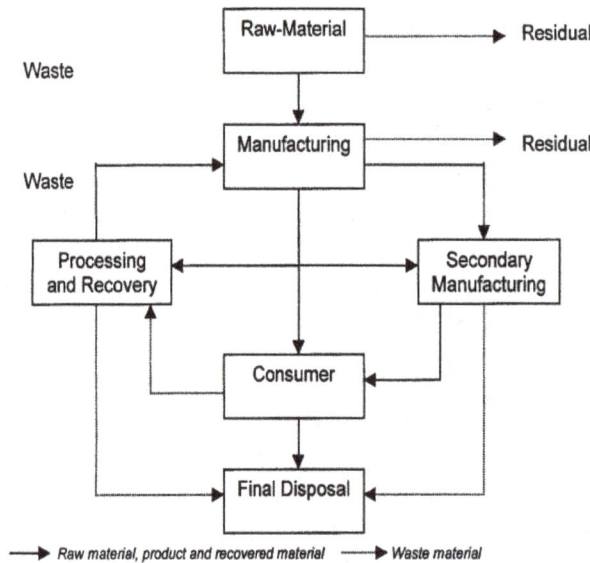

Material Flow and Waste

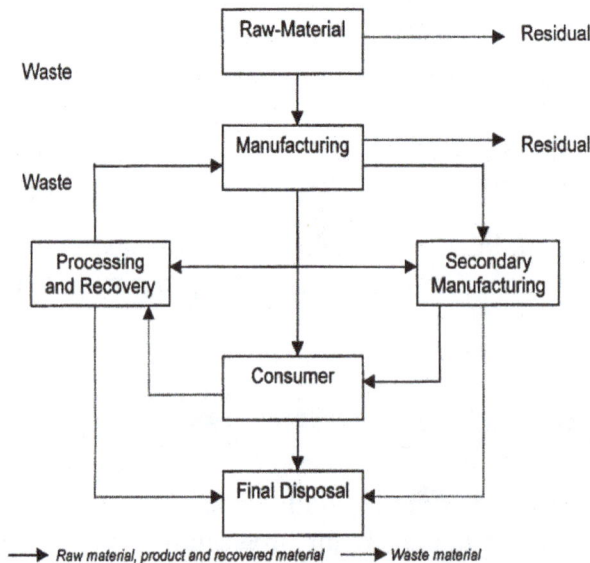

Generation

Figure suggests that we can reduce the amount of solid waste by limiting the consumption of raw materialsand increasing therate of recovery and reuse.

There needs to be, therefore, a societal change in the perception of wastes. This sounds simple. But, implementing changes in the society is difficult, unless appropriate management solutions are provided. That said, we also must note that the changes in waste generation contribute to changesinwaste composition.

Waste Composition

Some of the general observations associated with the composition of wastes include the following:

- The major constituents are paper and decomposable organic materials.

- More often than not, metal, glass, ceramics, textile, dirt and wood form part of the composition, and their relative proportion depends on local factors.

- Average proportions of the constituents reaching the disposal sites are consistent and urban wastes are fairly constant although subject to long-term changes such as seasonal variations.

Waste composition varies with the socio-economic status within a particular community, since income, for example, determines life style, composition pattern and cultural behaviour. Table illustrates this phenomenon in India:

Typical Waste Composition: Low/High Income Population

Characteristics	Low income	High income	Comments
Paper	1 – 4%	20 – 50%	Low paper content indicates low calorific value.
Plastics	1 – 6%	5 – 10%	Plastic is low as compared to high-income areas though the use of plastic has increased in recent years.
Ash and Fines	17 – 62%	3 – 10%	Ash and fines do not contribute to combustion process.
MoistureContent	30 – 40%	15 – 30%	Moisture content depends largely on the nature of the waste, climate and collection frequency. Waste can dry out while awaiting collection.
BulkDensity	300 – 400 kg/m^3	150 kg/m^3	Heavier waste may cost more to handle and difficult to burn.

Waste composition also depends on the moisture content, density and relative distribution of municipal wastes, as shown in Table below, and is important for the characterisation of solid waste for most applications (Ali, et al., 1999) :

Solid Wastes: Typical Composition, Moisture and Density

Components	Mass%		Moisture content %		Density in kg/m	
	Range	Typical	Range	Typical	Range	Typical
Food wastes	6-26	14	50-80	70	120-480	290
Paper	15-45	34	4-10	6	30-130	85
Cardboard	3-15	7	4-8	5	30-80	50
Plastics	2-8	5	1-4	2	30-130	65
Textiles	0-4	2	6-15	10	30-100	65
Rubber	0-1	0.5	1-4	2	90-200	130
Leather	0-2	0.5	8-12	10	90-260	160
Garden Trimming	0-20	12	30-80	60	60-225	105
Wood	1-4	2	15-40	20	120-320	240
Misc. Organic substances	0-5	2	10-60	25	90-360	240
Glass	4-16	8	1-4	2	160-480	195
Tincans	2-8	6	2-4	3	45-160	90
Non-ferrous metals	0-1	1	2-4	2	60-240	160
Ferrous metals	1-4	2	2-6	3	120-1200	320
Dirt, ash, bricks, etc.	0-10	4	6-12	8	320-960	480

Note that the density of waste changes as it moves from the source of generation to the point of ultimate disposal, and such factors as storage methods, salvaging activities, exposure to weather, handling methods and decomposition influence the density. In short, predicting changes of waste composition is as difficult as forecasting waste quantities.

Factors Causing Variation

As we know, wastes cause pollution. While the nature of wastes determines the type and intensity of pollution, it also helps us decide on the appropriate application, engineering design and technology for management. For example, the nature of wastes has implications for collection, transport and recycling. For effective SWM, therefore, we not only need information about the present but also the expected future quantity and composition of wastes.

There are several factors, which affect the present as well as the future waste quantity and composition (Tchobanoglous, et al., 1977), and some of which are listed below:

- Geographic location: The influence of geographic location is related primarily to different climates that can influence both the amount of certain types of solid wastes generated and the collection operation. For instance, substantial variations in the amount of yard and garden wastes generated in various parts of India are related to the climate. To illustrate, in the warmer southern areas, where the growing season is considerably longer compared to the northern areas, yard wastes are collected in considerably larger quantities and over a longer period of time.

- Seasons: Seasons of the year have implications for the quantities and composition of certain types of solid wastes. For example, the growing season of vegetables and fruits affect the quantities of food wastes.

- Collection frequency: A general observation is that in localities, where there are ultimate collection services, more wastes are collected. Note that this does not mean that more wastes are generated. For example, if a homeowner has access to only one or two containers perweek,dueto limited container capacity, he or she will store newspapers or other materials in some specified storage area. However, the same homeowner will tend to throw them away, if there is access to unlimited container services. In this latter situation, the quantity of waste generated may actually be the same but the quantity collected, as it relates to the frequency of collection, is considerably different.

- Population diversity: The characteristics of the population influence the quantity and composition of waste generated.Theamountofwaste generated is more in low-income areas compared to thatinhigh-income areas. Similarly, the composition differs in terms of paper and other recyclables, which are typically more in high-income areas as against low- income areas.

- Extent of salvaging and recycling: The existence of salvaging and recycling operation within a community definitely affects the quantity of wastes collected.

- Public attitude: Significant reduction in the quantity of solid waste is possible, if and when people are willing to change – on their own volition – their habits and lifestyles to conserve the natural resources and to reduce the economic burden associated with the management of solid wastes.

- Legislation: This refers to the existence of local and state regulations concerning the use and disposal of specific materials and is an important factor that influences the composition and generation of certain types of wastes. The Indian legislation dealing with packing and beverage container materials is an example.

In short elements that relate to waste generation include land use characteristics, population in age distribution, legislation, socio economic conditions, household and approximate number.

Waste Characterisation

Waste characterisation (*waste characterization* US) is the process by which the composition of different waste streams is analysed. Waste characterisation plays an important part in any treatment of waste which may occur. Developers of new waste technologies must take into account what exactly waste streams consist of in order to fully treat the waste. The biodegradable element of the waste stream is vitally important in the use of systems such as composting or anaerobic digestion.

Municipal waste streams are commonly broken down into the following constituents:

- Film plastic-LDPE

- Dense plastic-HDPE, PET

- Ferrous metal

- Non-ferrous metals

- Glass

- Textiles

- "Other" any remaining items which do not fit

Biodegradable fraction

- Glass

- Paper & cardboard

- Garden waste or green waste

- Fines (items below a certain screen size)

The European Waste Catalogue

Overview

The European Waste Catalogue (acronym EWC) refers to a set (although non-exhaustive) list of wastes that are derived from both households and businesses inside the European Union. The EWC is used to derive a code (six numbers in 3 sets of 2) that adequately describes the waste being transported, handled or treated. The EWC is used where Duty of Care Notices or Waste Transfer Notes are passed between waste management companies, waste carriers and to report volumes received or treated back to the governing agency (such as the Environment Agency in England and Wales, Scottish Environmental Protection Agency (SEPA) in Scotland, Northern Ireland (NI) Environment Agency, etc.).

Typical Waste Characterisation & Reporting

The first step in characterising waste is to decide on the appropriate EWC code. These codes carry three categories - absolute non-hazardous, mirror entries and absolute hazardous. Initial assessment for the majority of wastes follows a simple derivation of industry (some 20 main categories) from which they were originally obtained (Agricultural, wood working (furniture), Electronics, etc.). Each Derived section (denoted by the first 2 sets of 2 numbers refers to a particular industry or sector). The final set of 2 numbers relates directly to the waste. Where a waste is hazardous by its very composition, the EWC is followed by an Asterix).

Each member states Environment Agency throughout the European Union is obligated to adopt the EWC in its reporting methods and to enforce its use by the respective waste management sector. Submissions by Waste Management Companies to their respective Member states Environmental Agency are collated, in many instances by conversion to EWC STAT (European Waste Catalogue for Statistics) for submission to the EU, who oversees all member states and ensures compliance with unilaterally agreed standards and recycling rates.

Absolute Hazardous Wastes

Absolute hazardous entries are hazardous not due to the composition of the waste but by virtue of the process that produced it, the same is true for non-hazardous absolute entries. Mirror entries can either be hazardous or non-hazardous depending on the composition of the waste.

Deciding whether a mirror entry is hazardous or non-hazardous by composition involves reference to the Approved Supply List (ASL) initially. If a substance is not listed in the ASL then the regulations permit the use of other sources such as Manufacturer's Safety Data Sheets (MSDS) to classify the waste. These documents contain Risk Phrases, which describe the hazards that the substance or substances present. Risk phrases have threshold values attached to them that indicate what concentration of a substance must be present in order for the waste to be classified as hazardous by the Hazard Code attached to the Risk Phrase.

In order to identify the exact characteristics of municipal wastes, it is necessary that we analyse them using physical and chemical parameters (Phelps, et al., 1995).

Physical Characteristics

Information and data on the physical characteristics of solid wastes are important for the selection and operation of equipment and for the analysis and design of disposal facilities. The required information and data include the following:

(i) Density: Density of waste, i.e., its mass per unit volume (kg/m^3), is a critical factor in the design of a SWM system, e.g., the design of sanitary landfills, storage, types of collection and transport vehicles, etc. To explain, an efficient operation of a landfill demands compaction of wastes to optimum density. Any normal compaction equipment can achieve reduction in volume of wastes by 75%, which increases an initial density of 100 kg/m 3 to 400 kg/m^3. In other words, a waste collection vehicle can haul four times the weight of waste in its compacted state than when it is uncompacted. A high initial density of waste precludes the achievementofahigh compaction ratio and the compaction ratio achievedisnogreaterthan 1.5:1. Significant changes in density occur spontaneously as the waste moves from source to disposal, due to scavenging, handling, wetting and drying by the weather, vibration in the collection vehicle and decomposition. Note that:

- the effect of increasing the moisture content of the waste is detrimental in the sense that dry density decreases at higher moisture levels;

- soil-cover plays an important role in containing the waste;

- there is an upper limit to the density, and the conservative estimate of in-place density for waste in a sanitary landfill is about 600 kg/m^3.

(ii) Moisture content: Moisture content is defined as the ratio of the weight of water (wet weight - dry weight) to the total weight of the wet waste. Moisture increases the weight of solid wastes, and thereby, the cost of collection and transport. In addition, moisture content isacritical determinant in the economic feasibility of waste treatment by incineration, because wet waste consumes energy for evaporation of waterandin raising the temperature of water vapour. In the main, wastes should be insulated from rainfall or other extraneous water. We can calculate the moisture percentage, using the formula given below:

$$Moisture\,content\,(\%)=\frac{Wet weight - Dry weight}{Wet weight} \times 100$$

A typical range of moisture content is 20 to 40%, representing the extremes of wastes in an arid climate and in the wet season of a region of high precipitation.However, values greater than 40% are not uncommon.

(iii) Size: Measurement of size distribution of particles in waste stream is important because of its significance in the design of mechanical separators and shredders. Generally, the results of size distribution analysis are expressed in the manner used for soil particle analysis. That is to say, they are expressed as a plot of particle size (mm) against percentage, less than a given value.

The physical properties that are essential to analyse wastes disposed at landfills are:

I. Field capacity: The field capacity of MSW is the total amount of moisture which can be retained in a waste sample subject to gravitational pull. It is a critical measure because water in excess of field capacity will form leachate, and leachate can be a major problem in landfills. Field capacity varies with the degree of applied pressure and the state of decomposition of the wastes.

II. Permeability of compacted wastes: The hydraulic conductivity of compacted wastes is an important physical property because it governs the movement of liquids and gases in a landfill. Permeability depends on the other properties of the solid material include pore size distribution, surface area and porosity.

Porosity: It represents the amount of voids per unit overall volume of material. The porosity of MSW varies typically from0.40 to 0.67 depending on the compaction and composition of the waste.

Porosity of solid waste n= e/ (1+e) Where e is void ratio of solid waste

III. Compressibility of MSW: Degree of physical changes of the suspended solids or filter cake when subjected to pressure.

ΔHT =ΔHi +ΔHc +ΔHα [ΔHT= total settlement; ΔHi=immediate settlement; ΔHc = consolidation settlement; ΔHα =secondary compression or creep.]

C'α=ΔH/ [Ho X (Log (t2/t1))] = Cα/ (1+e0)

[Cα, C'α = Secondary compression index and Modified secondary Compression index; and t1, t2= Starting and ending time of secondary settlement respectively.]

Chemical Characteristics

Knowledge of the classification of chemical compounds and their characteristics is essential for the proper understanding of the behaviour of waste, as it moves through the waste management system. The products of decomposition and heating values are two examples of chemical characteristics. If solid wastes are to be used as fuel, or are used for any other purpose, we must know their chemical characteristics, including the following:

(i) Lipids: This class of compounds includes fats, oils and grease, and the principal sources of lipids are garbage, cooking oils and fats. Lipids have high heating values, about 38,000 kJ/kg (kilojoules per kilogram), which makes waste with high lipid content suitable for energy recovery. Since lipids become liquid at temperatures slightly above ambient, they add to the liquid content during wastedecomposition.Thoughtheyare biodegradable, the rate of biodegradation is relatively slow because lipids have a low solubility in water.

(ii) Carbohydrates: These are found primarily in food and yard wastes, which encompass sugar and polymer of sugars (e.g., starch, cellulose, etc.) with general formula $(CH_2O)x$. Carbohydrates are readily biodegraded to products such as carbon dioxide, water and methane. Decomposing carbohydrates attract flies and rats, and therefore, should not be left exposed for long duration.

(iii) Proteins: These are compounds containing carbon, hydrogen, oxygen and nitrogen, and consist of an organic acid with a substituted amine group (NH_2). They are mainly found in food and garden wastes. The partial decomposition of these compounds can result in the production of amines that have unpleasant odours.

(iv) Natural fibres: These are found in paper products, food and yard wastes and include the natural compounds, cellulose and lignin, that are resistant to biodegradation. (Note that paper is almost 100% cellulose, cotton over 95% and wood products over 40%.) Because they are a highly combustible solid waste, having a high proportion of paper and wood products, they are suitable for incineration. Calorific values of oven-dried paper products are in the range of 12,000 -18,000 kJ/kg and of wood about 20,000 kJ/kg, i.e., about half that for fuel oil, which is 44,200 kJ/kg.

(v) Synthetic organic material (Plastics) : Accounting for 1 – 10%, plastics have become a significant component of solid waste in recent years. They are highly resistant to biodegradation and, therefore, are objectionable and of special concern in SWM. Hence the increasing attention being paid to the recycling of plastics to reduce the proportion of this waste component at disposal sites. Plastics have a high heating value, about 32,000 kJ/kg, which makes them very suitable for incineration. But, you must note that polyvinyl chloride (PVC), when burnt, produces dioxin and acid gas. The latter increases corrosion in the combustion system and is responsible for acid rain.

(vi) Non-combustibles: This class includes glass, ceramics, metals, dust and ashes, and accounts for 12 – 25% of dry solids.

(vii) Heatingvalue: An evaluation of the potential of waste material for use as fuel for incineration requires a determination of its heating value, expressed as kilojoules per kilogram (kJ/kg). The heating value is determined experimentally using the Bomb calorimeter test, in which the heat generated, at a constant temperature of 25 C from the combustion of a dry sample is measured. Since the test temperature is below the boiling point ofwater(100 C), the combustion water remains in the liquid state.

However, during combustion, the temperature of the combustion gases reaches above 100 C, and the resultant water is in the vapour form. Table shows the typical inert residue and heating values for the components of municipal solid waste (Tchobanoglous, et al., 1977) :

Typical Heating and Inert Residue Values

Component	Inert Residue%		Heating Value (kJ/kg)	
	Range	Typical	Range	Typical
Food wastes	2-8	5	3500-7000	4500
Paper	4-8	6	11500-18500	16500
Card board	3-6	5	14000-17500	16000
Plastics	2-20	10	28000-37000	32500
Textiles	2-4	2.5	15000-20000	17500
Rubber	8-20	10	21000-28000	18500
Leather	8-20	10	15000-20000	17500
Garden trimmings	2-6	4.5	2300-18500	6500
Wood	0.6-2	1.5	17500-20000	18500
Glass	96-99	98	120-240	140
Tin cans	96-99	96	-	-
Nonferrous metals	90-99	96	240-1200	700
Ferrous metals	94-99	98	240-1200	700
Dirt, ash, bricks, etc.	60-80	70	2300-11500	7000
Municipal solid waste			9500-13000	10500

Note that while evaluating incineration as a means of disposal or energy recovery, we need to consider the heating values of respective constituents. For example:

- Organic material yields energy only when dry.

- The moisture content in the waste reduces the dry organic material per kilogram of waste and requires a significant amount of energyfor drying.

- The ash content of the waste reduces the proportion of dry organic material per kilogram of waste and retains some heat when removed from the furnace.

(viii) Ultimate analysis: This refers to an analysis of waste to determine the proportion of carbon, hydrogen, oxygen, nitrogen and sulphur, and the analysis is done to make mass balance calculation for a chemicalor thermal process. Besides, it is necessary to determine ash fraction because of its potentially harmful environmental effects, brought about by the presence of toxic metals such ascadmium, chromium, mercury,nickel, lead, tin and zinc. Note that other metals (e.g., iron, magnesium, etc.) may also be present but they are non-toxic. Table shows the resultof ultimate analysis of a typical municipal solid waste:

Municipal Solid Waste: A Typical Ultimate Analysis

Element	Range (% dry weight)
Carbon	25-30
Hydrogen	2.5-6.0
Oxygen	15-30
Nitrogen	0.25-1.2
Sulphur	0.02-0.12
Ash	12-30

(ix) Proximate analysis: This is important in evaluating the combustion properties of wastes or a waste or refuse derived fuel. The fractions of interest are:

- moisture content, which adds weight to the waste without increasing its heating value, and the evaporation of water reduces the heat released from the fuel;

- ash, which adds weight without generating any heat during combustion;

- volatile matter, i.e., that portion of the waste that is converted to gases before and during combustion;

- fixed carbon, which represents the carbon remaining on the surface grates as charcoal. A waste or fuel with a high proportion of fixed carbon requires a longer retention time on the furnace grates to achieve complete combustion than a waste or fuel with a low proportion of fixed carbon.

Table illustrates a proximate analysis for the combustible components of municipal solid waste:

Municipal Solid Waste: A Typical Proximate Analysis

Components	Value, percent	
	Range	Typical
Moisture	15-40	20
Volatilematter	40-60	53
Fixedcarbon	5-12	7
Glass,metal,ash	15-30	20

What subsection implies is that to evaluate alternative processing and recovery options (e.g., incineration process), we need information on the chemical characteristics of wastes, and wastes can typically be a combination of combustible and non-combustible materials.

Waste mismanagement, e.g., the practice of throwing wastes into streets, storm water drains, vacant land, etc., leads to breeding of disease vectors such as rats, with their attendant fleas carrying germs, etc., which results in an epidemic such as plague, malaria, etc. This has an adverse impact on public health and environment.

Solid Waste Collection

Solid waste collection does not merely mean collecting waste. Some of the aspects considered while collecting waste are public health and safety, collection route and storage containers. A transfer station is also necessary to act as a site of waste disposal. The aspects elucidated in this chapter are of vital importance, and provide a better understanding of solid waste collection.

Collection Components

Waste collection does not mean merely the gathering of wastes, and the process includes, as well, the transporting of wastes to transfer stations and/or disposal sites. To elaborate, the factors that influence the waste collection system include the following (EPA, 1989 and Ali, et al., 1999):

(i) Collection points: These affect such collection system componentsas crew size and storage, which ultimately control the cost of collection. Note that the collection points depend on locality and may be residential, commercial or industrial.

(ii) Collection frequency: Climatic conditions and requirements of a locality aswell as containers and costs determine the collection frequency. In hot and humid climates, for example, solid wastes must be collected at least twice a week, as the decomposing solid wastes produce bad odour and leachate.

And, as residential wastes usually contain food wastesandother putrescible (rotting) material, frequent collection is desirable for health and aesthetic reasons. Besides climates, the quality of solid waste containers on site also determines the collection frequency. For instance, while sealed or closed containers allow collection frequency up to three days, open and unsealed containers may require daily collection. Collection efficiency largely depends on the demography of the area (such as income groups, community, etc.), where collection takes place. While deciding collection frequency, therefore, you must consider the following:

- cost, e.g., optimal collection frequency reduces the cost as it involves fewer trucks, employees and reduction in total route distance;

- storage space, e.g., less frequent collection may require more storage space in the locality;

- sanitation, e.g., frequentcollectionreducesconcernsabouthealth, safety and nuisance associated with stored refuse.

(iii) Storage containers: Proper container selection can save collection energy, increase the speed of collection and reduce crew size. Most importantly, containers should be functional for the amount and type of materials and collection vehicles used.Containers should also be durable, easy to handle, and economical, as well as resistant to corrosion, weather and animals. In residential areas, where refuse is collected manually, standardised metal or plastic containers are typically required for waste storage. When mechanised collection systems are used, containers are specifically designed to fit the truck-mounted loading mechanisms. While evaluating residential wastecontainers, consider the following:

- efficiency, i.e., thecontainersshouldhelpmaximisetheoverall collection efficiency.

- convenience, i.e., the containers must be easily manageable both for residents and collection crew.

- compatibility, i.e., thecontainersmustbecompatiblewithcollection equipment.

- public health and safety, i.e., the containers should be securely covered and stored.

- ownership, i.e., the municipal ownership must guarantee compatibility with collection equipment.

(iv) Collection crew: The optimum crew size for a community depends on labour and equipment costs, collection methods and route characteristics. The size of the collection crew also depends on the size and type of collection vehicle used, space between the houses, waste generation rate and collection frequency. For example, increase in waste generation rate and quantity of wastes collected per stop due to less frequent collection result in a bigger crew size. Note also that the collection vehicle could be a motorised vehicle, a pushcart or a trailer towed by a suitable prime mover (tractor, etc.). It is possible to adjust the ratio of collectors to collection vehicles such that the crew idle time is minimised. However, it is not easy to implement this measure, as it may result in an overlap in the crew collection and truck idle time. An effective collection crew size and proper workforce management can influence the productivity of the collection system. The crew size, in essence, can have a great effect on overall collection costs. However, with increase in collection costs, the trend in recent years is towards:

- decrease in the frequency of collection.

- increase in the dependence on residents to sort waste materials; increase in the degree of automation used in collection.

- This trend has, in fact, contributed to smaller crews in municipalities.

(v) Collection route: The collection programme must consider the route that is efficient for collection. An efficient routing of collection vehicles helps decrease costs by reducing the labour expended for collection. Proper planning of collection route also helps conserve energy and minimise working hours and vehicle fuel consumption. It is necessary therefore to develop detailed route configurations and collection schedules for the selected collection system. The size of each route, however, depends on the amount of waste collected per stop, distance between stops, loading time and traffic conditions. Barriers, such as railroad, embankments, rivers and roads with heavy traffic, can be considered to divide route territories. Routing (network) analyses and planning can:

- increase the likelihood of all streets being serviced equally and consistently;

- help supervisors locate or track crews quickly;

- provide optimal routes that can be tested against driver judgement and experience.

(vi) Transfer station: A transfer station is an intermediate station between final disposal option and collection point in order to increase the efficiency of the system, as collection vehicles and crew remain closer to routes. If the disposal site is far from the collection area, it isjustifiable to have a transfer station, where smaller collection vehicles transfer their loads to larger vehicles, which then haul the waste long distances. In some instances, the transfer station serves as a pre- processing point, where wastes are dewatered, scooped or compressed. A centralised sorting and recovery of recyclable materials are also carried out at transfer stations (EPA, 1989). The unit cost of hauling solid wastes from a collection area to a transfer station and then to a disposal site decreases, as the size of the collection vehicle increases. This is duetovarious reasons such as the following:

- Labour costs remain constant;

- The ratio of payload to vehicle load increases with vehicle size;

- The waiting time, unloading time, idle time at traffic lights and driver rest period are constant, regardless of the collection vehicle size.

- Efficiency: Do the services help minimise the cost per household?

- Effectiveness: Do the services satisfy the community needs?

- Equity: Do the services address equally the concerns of all social and demographic groups?

- Reliability: Do the services ensure consistency?

- Safety and environmental impact: Do the services ensure safety of workers, public health and protection of the environment?

Note also that various management arrangements, ranging from municipal services to franchised services and under various forms of contracts are, typically, in vogue for waste collection. One of the critical decisions to be made at the planning stage, therefore, is as to who – the public or private agencies – operates the collection system, though the final decision depends on the existing conditions and options for the local decision-makers (EPA, 1989).

Waste Collection

Waste collection is a part of the process of waste management. It is the transfer of solid waste from the point of use and disposal to the point of treatment or landfill. Waste collection also includes the curbside collection of recyclable materials that technically are not waste, as part of a municipal landfill diversion program.

A waste collection vehicle in Sakon Nakhon, Thailand.

Manual waste collection in Bukit Batok West, Singapore.

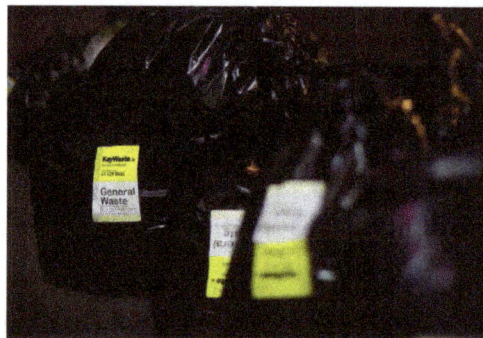

Waste on a sidewalk for collection, bagged and stickered - in Dublin, Ireland

Household Waste

Household waste in economically developed countries will generally be left in waste containers or recycling bins prior to collection by a waste collector using a waste collection vehicle.

However, in many developing countries, such as Mexico and Egypt, waste left in bins or bags at the side of the road will not be removed unless residents interact with the waste collectors.

Mexico City residents must haul their trash to a waste collection vehicle which makes frequent stops around each neighborhood. The waste collectors will indicate their readiness by ringing a distinctive bell and possibly shouting. Residents line up and hand their trash container to the waste collector. A tip may be expected in some neighborhoods. Private contracted waste collectors may circulate in the same neighborhoods as many as five times per day, pushing a cart with a waste container, ringing a bell and shouting to announce their presence. These private contractors are not paid a salary, and survive only on the tips they receive. Later, they meet up with a waste collection vehicle to deposit their accumulated waste.

The waste collection vehicle will often take the waste to a transfer station where it will be loaded up into a larger vehicle and sent to either a landfill or alternative waste treatment facility.

Commercial Waste

Waste collection considerations include type and size of bins, positioning of the bins, and how often bins are to be serviced. Overfilled bins result in rubbish falling out while being tipped. Hazardous rubbish like empty petrol cans can cause fires igniting other trash when the truck compactor is operating. Bins may be locked or stored in secure areas to avoid having non-paying parties placing rubbish in the bin.

Kerbside Collection

Kerbside collection in Canberra, Australia

Kerbside collection, orcurbside collection, is a service provided to households, typically in urban and suburban areas of removing household waste. It is usually accomplished by personnel using purpose built vehicles to pick up household waste in containers acceptable to or prescribed by the municipality.

History

Prior to the 20th century the amount of waste generated by a household was relatively small. Household wastes were often simply thrown out the window, buried in the garden or deposited in outhouses. When human concentrations became more dense, waste collectors, called nightmen or gong farmers were hired to collect the night soil from pail closets, performing their duties only at night (hence the name). Meanwhile, disposing of refuse became a problem wherever cities grew. Often refuse was placed in unusable areas just outside the city, such as wetlands and tidal zones. One example is London, which from Roman times disposed of its refuse outside the London Wall beside the River Thames. Another example is 1830s Manhattan, where thousands of hogs were permitted to roam the streets and eat garbage. A small industry developed as "swill children" collected kitchen refuse to sell for pig feed and the rag and bone man traded goods for bones (used for glue) and rags (essential for paper manufacture prior to the invention of wood pulping). Later, in the late nineteenth century, trash was fed to swine in industrial.

As sanitation engineering came to be practised beginning in the mid-19th century and human waste was conveyed from the home in pipes, the gong farmer was replaced by the municipal trash collector as there remained growing amounts of household refuse, including fly ash from coal, which was burnt for home heating. In Paris, the rag and bone man worked side by side with the municipal bin man, though reluctantly: in 1884, Eugène Poubelle introduced the first integrated kerbside collection and recycling system, requiring residents to separate their waste into perishable items, paper and cloth, and crockery and shells. He also established rules for how private collectors and city workers should cooperate and he developed standard dimensions for refuse containers: his name in France is now synonymous with the garbage can. Under Poubelle, food waste and other organics collected in Paris were transported to nearby Saint Ouen where they were composted. This continued well into the 20th century when plastics began to contaminate the waste stream.

From the late-19th century to the mid-20th century, more or less consistent with the rise of consumables and disposable products municipalities began to pass anti-dumping ordinances and introduce kerbside collection. Residents were required to use a variety of refuse containers to facilitate kerbside collection but the main type was a variation of Poubelle's metal garbage container. It was not until the late 1960s that the green bin bag was introduced by Glad. Later, as waste management practices were introduced with the aim of reducing landfill impacts, a range of container types, mostly made of durable plastic, came to be introduced to facilitate the proper diversion of the waste stream. Such containers include blue boxes, green bins and wheelie bins or MGBs.

Over time, waste collection vehicles gradually increased in size from the hand pushed tip cart or English dust cart, a name by which these vehicles are still referred, to large compactor trucks.

Waste Management and Resource Recovery

Glass for collection in Edinburgh, Scotland.

Kerbside collection is today often referred to as a strategy of local authorities to collect recyclable items from the consumer. Kerbside collection is considered a low-risk strategy to reduce waste volumes and increase recycling rates. Materials are typically collected in large bins, coloured bags, or small open plastic tubs, specifically designated for content.

Discarded Christmas trees awaiting pickup in the San Fernando Valley

Recyclable materials that may be separately collected from municipal waste include:

Biodegradable waste component

- Green waste

- Kitchen waste

Recyclable materials, depending on location

- Office paper

- Newsprint

- Paperboard

- Corrugated fiberboard

- Plastics (#1 PET, #2 HDPE natural and colored, #3 PVC narrow-necked containers, #4 LDPE, #5 PP, #6 Polystyrene (however not EXPANDED polystyrene, an example of recyclable polystyrene may be a yoghurt pot) #7 other mixed resin plastics)

- Glass

- Copper

- Aluminum

- Steel and Tinplate

- Co-mingled recyclables- can be sorted by a clean materials recovery facility

In Somerville, MA all accepted paper, glass, plastic, and metal
recycling is picked up from a single bin

Kerbside collection of recyclable resources is aimed to recover purer waste streams with higher market value than by other collection methods. If the household incorrectly separates the recyclable elements, the load may have to be put to landfill if it is deemed to be contaminated.

Kerbside collection and household recycling schemes are also being used as tools by local authorities to increase the public's awareness of their waste production.

Kerbside collection is commonly considered to be completely environmentally friendly. This may not necessarily be the case as it leads to an increased number of waste collection vehicles on the road, in themselves contributing to global warming through exhaust emissions until the time of their conversion to clean energy.

New and emerging waste treatment technologies such as mechanical biological treatment may offer an alternative to kerbside collection through automated separation of waste in recycling factories.

Usage by Country

Canada

Calgary, Alberta has adopted "Curbside" Recycling and uses blue bins. The blue cart programme accepts all types of recycables, including plastics 1-7. It is picked up weekly for the cost of $8.00 per month. This programme is mandatory.

In 1981 Resource Integration Systems (RIS) in collaboration with Laidlaw International tested the first blue box recycling system on 1500 homes in Kitchener, Ontario. Due to the success of the project the City of Kitchener put out a contract for public bid in 1984 for a recycling system city wide. Laidlaw won the bid and continued with the popular blue box recycling system. Today hundreds of cities around the world use the blue box system or a similar variation.

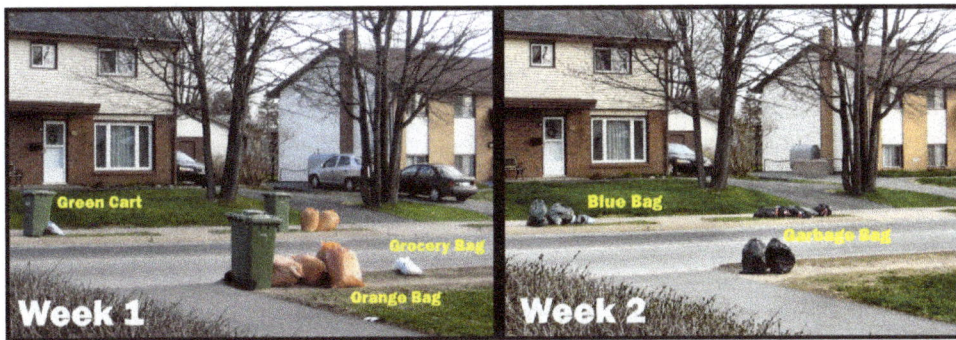

Halifax Regional Municipality (HRM) in Nova Scotia, Canada, with a population of about 375, 000, has one of the most complex kerbside collection programmes in North America. Based on the green cart, it requires residents to self-sort refuse and place different types at the kerb on alternating weeks. As shown in the photo at left, week 1 would see the green cart and optional orange bags used for kitchen waste and other organics such as yard waste. Week 2 would permit non-recoverable waste in garbage bags or cans. Blue bags are used for paper, plastic and metal containers. Together with used grocery bags containing newspapers, they may be placed on the kerb either week. In summer, the green cart is emptied weekly due to the prevalence of flies. HRM has achieved a diversion rate of approximately 60 percent by this method.

Many Canadian municipalities use "green bins" for kerbside recycling. Others, such as Moncton, use wet/dry waste separation and recovery programmes.

New Zealand

Kerbside collection bins in Dunedin, New Zealand. The yellow-liddied wheelie bin is for non-glass recyclables, and the blue bin is for glass. The two bins are collected on alternating weeks. Official council bags are used for general household waste, and are collected weekly.

In New Zealand, kerbside collection of general refuse and recycling, and in some areas organic waste, is the responsibility of the local city or district council, or private contractors. Practices and collection methods vary widely from council to council and company to company. Some examples of collection are:

- Auckland City Council: Two 240-litre wheelie bins are supplied: a red-lidded bin for general refuse, collected weekly, and a blue-lidded bin for recyclables, collected fortnightly.

- Christchurch City Council: Three wheelie bins are supplied: a 140-litre red-lidded bin for general refuse, a 240-litre yellow-lidded bin for recyclables, and an 80-litre green-lidded bin for organic waste. The organic waste bins are collected weekly, while the recyclables and general refuse bins are collected on alternating weeks.

- Hamilton City Council and Hutt City Council: A 45-litre bin is supplies for recyclables, collected weekly. General refuse is collected weekly using user-pays official council bags.

- Dunedin City Council, Palmerston North City Council and Wellington City Council: Two bins are supplied. A 45-litre or 70-litre bin for glass, and an 80-litre or 240-litre wheelie bin for non-glass recyclables. These two bins are collected on alternating weeks. General refuse is collected weekly using user-pays official council bags.

- Rodney District Council: A 45-litre bin is supplies for recyclables, collected weekly. There is no council collection of general waste, and all general waste collection is carried out by independent companies.

- Taupo District Council: A 45-litre bin is supplies for recyclables, collected weekly. General refuse is collected weekly using user-pays system of orange tags - one orange tag is to be placed on a standard rubbish bag up to 60 litres capacity, or half an orange sticker can be placed on two supermarket bags tied together.

- Upper Hutt City Council: Recycling is to be placed in plastic bags, with paper and cardboard collected in the first week, and plastic, metal and glass in the second week. General refuse is collected weekly using user-pays official council bags.

- Waitakere City Council: A 140-litre wheelie bin is provided for recyclables, collected fortnightly. General refuse is collected weekly using user-pays official council bags.

By 1996 the New Zealand cities of Auckland, Waitakere, North Shore and Lower Hutt had kerbside recycling bins available. In New Plymouth, Wanganui and Upper Hutt recyclable material was collected if placed in suitable bags. By 2007 73% of New Zealanders had access to kerbside recycling.

Kerbside collection of organic waste is carried out by the Mackenzie District Council and the Timaru District Council. Christchurch City Council is introducing the system to their kerbside collection. Other councils are carrying out trials.

United Kingdom

In the United Kingdom, the Household Waste Recycling Act 2003 requires local authorities to provide every household with a separate collection of at least two types of recyclable materials by 2010. There has been criticism in the difference of schemes used in the country such as the colour of bins, whether they are bins boxes or bags, and also the fact that clutter roads and how the additional trucks and collections needed have carbon dioxide emissions too. Some find the colour differences confusing, and people want a national scheme. A typical example is to compare two neighbouring councils in greater Manchester, Bury council and Salford. Bury uses blue for cans, plastic and glass, green for paper and cardboard and brown for garden waste. Salford uses blue for paper and card, brown for cans plastic and glass and pink for garden waste. Most councils use grey or black for general waste, with a few exceptions such as Liverpool, which uses purple for general waste, a colour used by no other council

Another controversial issue in the uk is the frequency of the waste collections. To save money, many councils are cutting the frequency of both general waste and recyclables collections. This has led to problems from larger families, and has led to overflowing and fly tipping. For example, previously, Bury Council collected general waste once a week and recyclables fortnightly. This has now changed to fortnightly for general waste and monthly (every 4 weeks) collection of recyclables.

A few councils are using "forced" recycling, by replacing the large, 240l general waste bin with a smaller 180l or 140l bin, and using the old 240l one for recyclables. This may be made worse by fortnightly collections of the "small" bin, and strict rules such as "No extra bags will be taken" and "Bin lids must be fully closed". Stockport Council is a notable user of this scheme. Their recycling rates have risen substantially as a result, but there are usually complaints from families. Trafford council also use a similar scheme, but the small grey bin is emptied every week. In addition, the two named councils, and more, collect food waste together with garden waste, by sending out kitchen caddies and compostable liners. These prevent food waste (including meat) from going to landfill, and to increase the councils recycling rate. The food and garden waste is usually collected weekly or fortnightly, and is taken to an In Vessel composter or Anaerobic digester, where the waste is turned into soil improver for use on local farms.

In the north west, all the glass collected is used within the UK, around half of the plastics and cans are used in the UK; the rest is sent further afield to Europe or China to be made into new products, and paper and paperboard collected is sent to local paper mills to be made into newspapers, tissues, paperboard and office paper. Again some of the paper will be sent further afield.

Some councils only have 3 bins- general, organic and recyclables. This means that plastics, cans and glass go in the same container as paper and cardboard. Although this is much easier for the residents, there is more sorting required, and the paper quality is sometimes of a low grade due to food contamination or shards of glass in the paper, and so this scheme is criticized.

Also, most councils require residents to remove caps from bottles and rinse them out to avoid smells. This is because the lids are made from a different type of plastic (PP) to the bottle (PET/HDPE) - although by collapsing the bottles and folding them over like toothpaste tubes and re-screwing the caps in place enables the volume of bottles to be drastically reduced, thereby increasing the amount of bottles that can be carried in the recycling bins. In fact many bottlers, especially bottled water companies, have now designed their bottles to be collapsible; though this message has not been effectively disseminated to the consumer. A collapsible bottle takes between 25% and 33% of the space a non-collapsed bottle.

Labels are rarely required to be removed, however. This also means that only plastic bottles are recycled. Councils are still trying to make clear that plastic tubs (yogurts, desserts and spreads), bags and cling film cannot be recycled through the kerbside economically. If too much contamination is collected then this results in the whole vehicle load going to landfill at a high cost. Contamination is usually a problem if recyclables are collected in wheelie bins, as the worker can only look at the top; there may be contamination 'hidden' at the bottom. Councils that use many bags and boxes (Edinburgh) suffer from less contamination but are complicated and the loose paper and cardboard, and recycling bags are blown around, and paper can be wet.

Basque Country

In the province of Gipuzkoa, this system is implanted in many towns as Usurbil, Hernani, Oiartzun, Antzuola, Legorreta, Itsasondo, Zaldibia, Anoeta, Alegia, Irura, Zizurkil, Astigarraga, Ordizia, Oñati and Lezo, where the common used name in basque is atez-atekoa, which means *door-by-door*. Due to the big success in this towns, with more than the 80% of the waste recycled, 34 towns in Gipuzkoa are studying to set this system up in 2013, like Arrasate, Bergara, Aretxabaleta, Eskoriatza, Legazpi, Tolosa or Pasaia.

The atez-ate system consists in hanging each kind of rubbish in a hanger outside the house a certain day or days in a week. For example, in Hernani, they have three days to hang their organic rubbish, two days for plastics and metallics, one for paper and one for rejects residuals.

This system started in the town of Usurbil in the year 2009, due to the incinerator of the region of Gipuzkoa was going to build in this town, exactly in the neighborhood of Zubieta. Three years after, the construction of the incinerator was paralyzed by the government of the region, suggesting that the incinerator was a source of contamination and the high cost of the building.

Criticism

This type of collection service is subject to growing criticism.

- The large (Wheelie bin) container encourages the "out of sight" rubbish mentality and invites more rubbish to be disposed.

- The bins and collection trucks are not suited to narrow roads or houses with steep driveways or steps.

- They lock local authorities into capital intensive equipment programmes and multi-national providers.

- Co-mingled recyclables are sometimes not being successfully managed by automated sorting stations and the rates of diversion are low. In some cases, this results in mountains of unsorted recyclables.

- In the UK especially, some councils are sending out at least 4 large bins - residents of smaller houses with no gardens have little space to put them

- Many use small plastic boxes, bags and lockable outdoor food waste 'caddies' which get blown around and lost, bad for recycling participation.

Collection Operation

Now lets discuss the movement of collection crew in terms of workforce efficiency and collection routes.

Movement of Collection Crew

In cultures such as India, Bangladesh, etc., solid waste collection is assigned to the lowest social group. More often, the collection crew member accepts the job as a temporary position or stopgap arrangement, while looking for other jobs that are considered more respectable.

Apart from this cultural problem, the attitude of some SWM authorities affects collection operation. For example, some authorities still think that the collection of solid waste is mechanical, and therefore, the collection crew does not need any training to acquire special skills. As a result, when a new waste collector starts working, he or she is sent to the field without firm instruction concerning his or her duties, responsibilities and required skills. For an effective collection operation, the collection team must properly be trained. The collection crew and the driver of the collection vehicle must, for example, work as a team, and this is important to maintain the team morale and a sense of social responsibility among these workers.

You must also note that the movement of collection crew, container location and vehicle stopping point affect collection system costs. Figure highlights the distance the collection crew will have to walk, if it were to serve the farthest point first or serve the point closest to the vehicle:

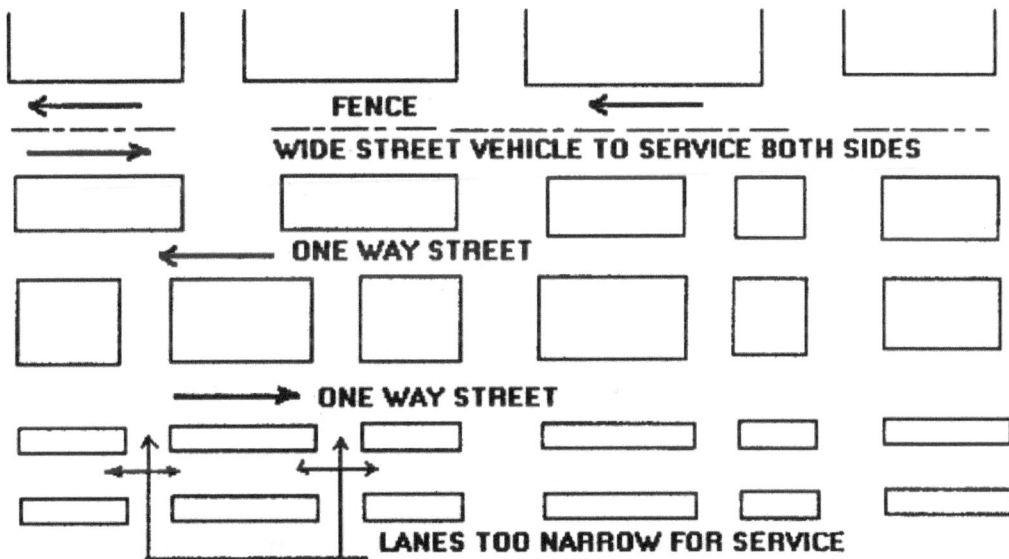

Effect of Container Location and Vehicle Stopping

The difference may be one or two minutes per collection stop, but it matters with the number of stops the crew will take in a working shift. Multiplying the minutes by the total number of crew working and labour cost depicts the amount of labour hours lost in terms of monetary value.

Generally, familiarity of the crew with the collection area improves efficiency. For example, the driver becomes familiar with the traffic jams, potholes and other obstructions that he or she must avoid. The crew is aware of the location of the containers and the vehicle stops. It is, therefore, important to assign each crew specific areas of responsibility. Working together also establishes an understanding of the strong and weak points of the team members and efficient work sequences. The collection operation must also observe a strict time schedule. Testing of new routes, new gadgets and vehicles is best carried out first in the laboratory and later in a pilot area. Testing of a new sequence using the whole service area could result in disorder and breakdown of the solid waste collection system. Studies show that it takes two hours to recover for every hour of a failed system.

Motion Time Measurement (MTM) Technique

Motion time measurement (MTM) studies are now an integral part of the standard procedure in the development of solid waste collection systems. MTM is a technique to observe and estimate the movement of the collection crew with the help of stopwatches.The results thus gathered are tabulated as shown in Table to determine the best sequence of activities that workers must follow in order to complete a repetitive task in the shortest possible time:

Table:

MTM Study: Determination of Time, Distance and Number on Containers in Collection Route

	Time		Odometer (Km)	Number of Containers	Collection time (Minute Second)	Trip time to next Station
	Arrival	Departure				
Garage	::	::				
1 Station	::	::				
2 Station	::	::				
.	::	::				
.	::	::				
.	::	::				
20 Station	::	::				
Last Station	::	::				
Disposal Site Total	::	::				
Weight	With Load tonne		With Load tonne		With Load tonne	

MTM also helps in deciding the best combination of equipment to maintain a desired level of output, reduce health problems related to the repetitive work sequence and predict the effects of changes in materials handled.

Sophisticated MTM studies involve hidden or open video cameras at different collection stops to record, replay and study the operation sequence of the collection crew. If the crew is conscious of being observed, they tend to work faster and reduce time wastage in unauthorised salvaging and other non- scheduled activities. Once the crew is familiar with the person(s) observing them, it begins to perform more credibly. In studies involving video cameras, therefore, the first two or three hours of observation are often neglected.

Collection Vehicle Routing

Efficient routing and re-routing of solid waste collection vehicles can help decrease costs by reducing the labour expended for collection. Routing procedures usually consist of the following two separate components:

(i) Macro-routing: Macro-routing, also referred to as route-balancing, consists of dividing the total collection area into routes, sized in such a way as to represent a day's collection for each crew. The size of each route depends on the amount of waste collected per stop, distance between stops, loading time and traffic conditions. Barriers, such as railroad embankments, rivers and roads with heavy competing traffic, can be used to divide route territories. As much as possible, the size and shape of route areas should be balanced within the limits imposed by such barriers.

(ii) Micro-routing: Using the results of the macro-routing analysis, micro- routing can define the specific path that each crew and collection vehicle will take each collection day. Results of micro-routing analyses can then be used to readjust macro-routing decisions. Micro-routinganalysesshould also include input and review from experienced collection drivers.

Districting is the other method for collection route design. For larger areas it is not possible for one institution to handle it then the best way is to sub divide the area and MSW collection districting plan can be made.This routing will be successful only when road network integrity is good and the regional proximity has been generated.

The heuristic (i.e., trial and error) route development processisarelatively simple manual approach that applies specific routing patterns to block configurations. The map should show collection, service garage locations, disposal or transfer sites, one-way streets, natural barriers and areas of heavy traffic flow. Routes should then betracedontothetracingpaperusingthe following rules:

- Routes should not be fragmented or overlapping. Each route should be compact, consisting of street segments clustered in the same geographical area.

- Total collection plus hauling time should be reasonably constant for each route in the community.

- The collection route should be started as close to the garage or motor pool as possible, taking into account heavily travelled and one-way streets.

- Heavily travelled streets should not be visited during rush hours.

- In the case of one-way streets, it is best to start the route near the upper end of the street, working down it through the looping process.

- Services on dead-end streets can be considered as services on the street segment that theyintersect, since theycan onlybe collected by passing down that street segment. To keep right turns at a minimum, (in countries where driving is left-oriented) collection from the dead-end streets is done when they are to the left of the truck. They must be collected by walking down, reversing the vehicle or taking a U-turn.

- Waste on a steep hill should be collected, when practical, on both sides of the street while vehicle is moving downhill. This facilitates safe, easy and fast collection. It also lessens wear of vehicle and conserves gas and oil.

- Higher elevations should be at the start of the route.

- For collection from one side of the street at a time, it is generally best to route with many anti-clockwise turns around blocks.

- For collection from both sides of the street at the same time, it is generally best to route with long, straight paths acrossthe grid before looping anti- clockwise.

- For certain block configurations within the route, specific routingpatterns should be applied. (Adapted from American Public Works Association, 1975.)

Based on the above rules, Figure below illustrates a typical collection vehicle routing:

Figure: Collection Vehicle Route

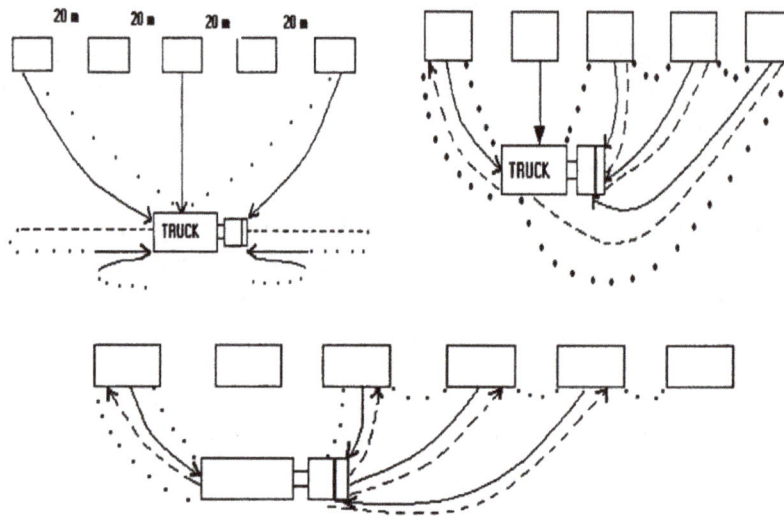

Waste Collection System Design

After we identify appropriate options for collection, equipment and transfer, we must examine the various combinations of these elements to define system -wide alternatives for further analysis. Each should be evaluated for its ability to achieve the identified goals of the collection programme. Economic analysis will usually be a central focus of the system evaluation. This initial evaluation will lead to several iterations, with the differences between the alternatives under consideration becoming more narrowly focused with each round of evaluations (EPA, 1995). After comparing the alternative strategies, the various elements like crew and truck requirement, time requirement and cost involved are calculated. The various formulae used to calculate are:

(i) Number of services/vehicle load (N):

$N = (C x D) / W$

where, C = Vehicle capacity (m³) ; D = Waste density (kg/m³) and W = Waste generation/residence (kg/service)

(ii) Time required collecting one load (E):

$E = N \, x \, L$

where, L = Loading time/residence, including on-route travel

(iii) Number of loads/crew/day (n):

The number of loads (n) that eachcrew cancollect in a day canbe estimated based on the workday length (t), and the time spent on administration and breaks (t_1), time for hauling and other travel (t_2) and collection route time (t_3).

Administrative and break time (t_1):

$t_1 = A + B$

where, A=Administrativetime(i.e., formeetings, paperwork, unspecified slack time) and B = Time forbreaks and lunch

Hauling and other travel time (t_2):

$t_2 = (n \, x \, H) - f + G + J$

where, n = Number of loads/crew/day; H = Time to travel to disposal site, empty truck, and return to route; f = Time to return from site to route; G = Time to travel from staging garage to route and J = Time to return from disposal site to garage.

Time spent on collection route (t_3):

$t_3 = n \, x \, E$

where variables have been previously defined.

Length of workday (t):

$t = t_1 + t_2 + t_3$

where t is defined by work rules and equations A through D are solved to find n.

(iv) Calculation of number of vehicles and crews (K):

$K = (S \, x \, F) / (N \, x \, n \, x \, M)$

where, S = Total number of services in the collection area; F = Frequency of collection (numbers/week) and M =Number of workdays/week

(v) Calculation of annual vehicle and labour costs: Vehicle costs =Depreciation + Maintenance + Consumables + Overhead + License + Fees + Insurance

Labour costs =Drivers salary + Crew salaries + Fringe benefits + Indirect labour + Supplies + Overhead

Storage: Containers/Collection Vehicles

Waste storage encompasses proper containers to store wastes and efficient transport of wastes without any spillage to transfer stations/disposal sites.

Containers/Storage Bins

The design of an efficient waste collection system requires careful consideration of the type, size and location of containers at the point of generation for storage of wastes until they are collected. While single-family households generally use small containers, residential units, commercial units, institutions and industries require large containers. Smaller containers are usually handled-manually whereas the larger, heavier ones require mechanical handling. The containers may fall under either of the following two categories:

(i) Stationary containers: These are used for contents to be transferred to collection vehicles at the site of storage.

(ii) Hauled containers: These are used for contents to be directly transferred to a processing plant, transfer station or disposal site for emptying before being returned to the storage site.

The desirable characteristics of a well-designed container are low cost, size, weight, shape, resistance to corrosion, water tightness, strength and durability (Phelps, et al., 1995). For example, a container for manual handling by one person should not weigh more than 20 kg, lest it may lead to occupational health hazards such as muscular strain, etc. Containers that weigh more than 20 kg, when full, require two or more crew members to manually load and unload the wastes, and which result in low collection efficiency.

Containers should not have rough or sharp edges, and preferably have a handle and a wheel to facilitate mobility. They should be covered to prevent rainwater from entering (which increases the weight and rate of decomposition of organic materials) into the solid wastes. The container body must be strong enough to resist and discourage stray animals and scavengers from ripping it as well as with stand rough handling by the collection crew and mechanical loading equipment. Containers should be provided with a lifting bar, compatible with the hoisting mechanism of the vehicle. The material used should be light, recyclable, easily moulded and the surface must be smooth and resistant to corrosion. On the one hand, steel and ferrous containers are heavy and subject to corrosion; the rust peels off exposing sharp edges, which could be hazardous to the collection crew. On the other, wooden containers (e.g., bamboo, rattanand wooden baskets) readily absorb and retain moisture and their surfaces are generally rough, irregular and difficult to clean.

Communal Containers

Generally, the containers used for waste storage arecommunal/public containers. Figure below shows a typical communal container, which a compactor collection vehicle can lift and empty mechanically:

The use of communal containers is largely dependent on local culture, tradition and attitudes towards waste. Communal containers may be fixed on the ground (stationary) or movable (hauled).

Movable containers are provided with hoists and tails compatible with lifting mechanism of collection vehicles and such containers have capacities of 1 – 4 m3. The waste management authority must monitor, maintain and upgrade the communal containers. Note that in residential and commercial areas in India, the communal containers are often made of concrete.

Typical Communal Container

In areas with very high waste generation rates, i.e., rates exceeding two truckloads daily, such as wet markets, large commercial centres and large business establishments, roll-on-roll or hoisted communal containers with capacities of 12 – 20 m3 and a strong superstructure with wheels are used. Normally, the collection vehicle keeps an empty container asareplacement before it hauls the filled container. When a truck is used as a collection vehicle, the use of communal containers may be appropriate.

It is advisable to place the containers 100 – 200 m apart for economic reasons. The communal containers are usually staggered such that the effective distance of 100 m is maintained as shown in Figure:

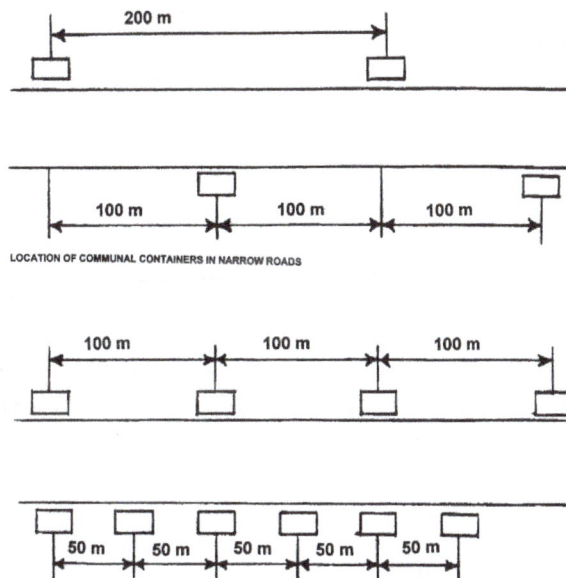

Location of Communal Container

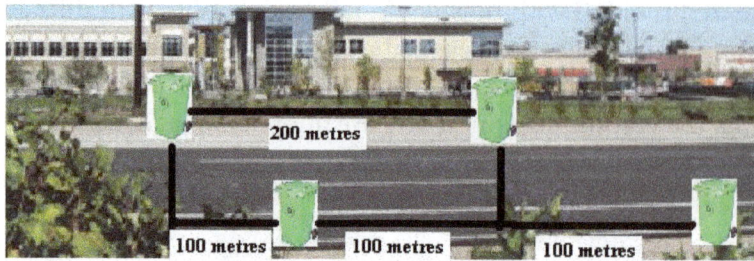

Location of communal containers in Narrow Roads

200 metres

100 metres | 100 metres | 100 metres

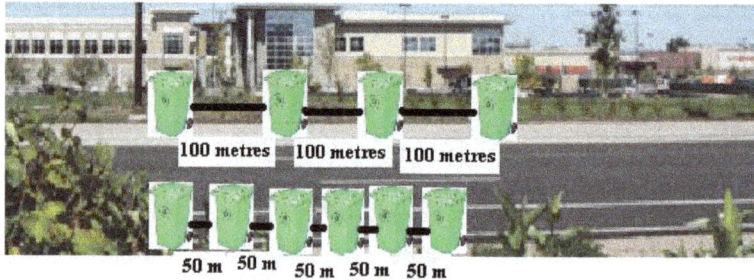

100 metres | 100 metres | 100 metres

50 m | 50 m | 50 m | 50 m | 50 m

This means that the farthest distance the householder will have to walk is 50 meters. However, in narrow streets with low traffic, where the house owner can readily cross the street, a longer distance is advisable. If the collection vehicle has to stop frequently, say, at every 50 m or so, fuel consumption increases, and this must be avoided.

Disadvantages

The major disadvantage of communal containers is the potential lack of maintenance and upgrading. The residuals and scattered solid wastes emit foul odours, which discourage residents from using the containers properly. In addition, if fixed containers are built below the vehicle level, the collection crew may be held responsible for sweeping and loading the solid wastes into transfer containers before being loaded into the collection vehicle. Sweepingand cleaning the communal containers of residuals obviously impinge on the time of the crew members and take a longer time than if the wastes are placed in smaller containers. As fixed communal containers have higher rates of failure, their use is not advisable.

To overcome the problem of maintaining communal containers, individual residents should maintain their own containers and locate them in designated areas. The communal area must have water and drains to facilitate the cleaning of the containers. This practice has the advantage of reducing the number of collection stops and at the same time maintaining the householder's responsibility for cleaning them. The residents must also be properly educated on the importance of good housekeeping as the containers in the communal area are subject to vandalism. In the main, if communal containers are to be successful, the design of the containers, loading and unloading areas, and collection vehicle accessories should be co-ordinated.

Collection Vehicles

Almost all collections are based on collector and collection crew, which move through the collection service area with a vehicle for collecting the waste material. The collection vehicle selected must be appropriate to the terrain, type and density of waste generation points, the way it travels and type and kind of material (UNEP, 1996). It also depends upon strength, stature and capability of the crew that will work with it. The collection vehicle may be small and simple (e.g., two-wheeled cart pulled by an individual) or large, complex and energy intensive (e.g., rear loading compactor truck). The most commonly used collection vehicle is the dump truck fitted with a hydraulic lifting mechanism. A description of some vehicle types follows:

(i) Small-scale collection and muscle-powered vehicles: These are common vehicles used for waste collection in manycountries and are generally used in rural hilly areas. As Figure illustrates, these can be small rickshaws, carts or wagons pulled by people or animals, and are less expensive, easier to build and maintain compared to other vehicles:

They are suitable for denselypopulated areas with narrow lanes, and squatter settlements, where there is relatively low volume of waste generated. Some drawbacks of these collection vehicles includelimited travel range of the vehicles and weather exposure that affect humans and animals.

Small-scale Collection Vehicles: An Illustration

(ii) Non-compactor trucks: Non-compactor trucks are efficient and cost effective in small cities and in areas where wastes tend to be very dense and have little potential for compaction. Figure illustrates a non- compactor truck:

Non-compactor Trucks

When these trucks are used for waste collection, they need a dumping system to easily discharge the waste. It is generally required to cover the trucks in order to prevent residue flying off or rain soaking the wastes. Trucks with capacities of 10 – 12 m3 are effective, if the distance between the disposal site and the collection area is less than 15 km. If the distance is longer, a potential transfer station closer than 10 km from the collection area is required. Non-compactor trucks are generally used, when labour cost is high. Controlling and operating cost is a deciding factor, when collection routes are long and relatively sparsely populated.

(iii) Compactor truck: Compaction vehicles are more common these days, generally having capacities of 12 – 15 m3 due to limitations imposed by narrow roads. Although the capacity of a compaction vehicle, illustrated in Figure, is similar to that of a dump truck, the weight of solid wastes collected per trip is 2 to 2.5 times larger since the wastes are hydraulically compacted:

Compactor Truck

The success of waste management depends on the level of segregationat source. One of the examples for best collection method is illustrated in the figure below:

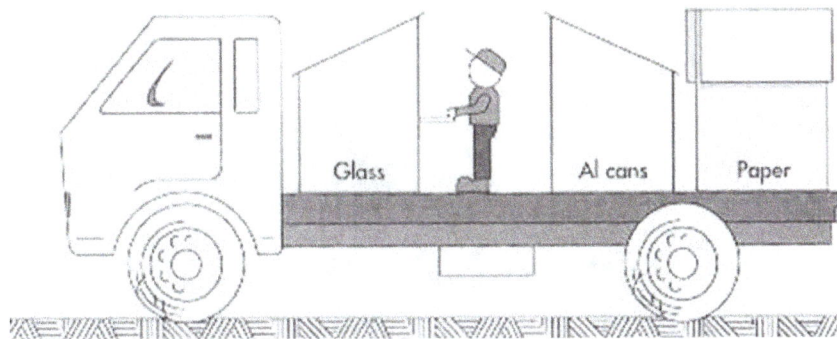

A compactor truck allows waste containers to be emptied into the vehicle from the rear, front or sides and inhibits vectors (of disease) from reaching the waste during collection and transport. It works poorly when wastestream is very dense, wet, collected materials are gritty or abrasive, or when the roads are dusty. The advantages of the compactor collection vehicle include the following:

- containers are uniform, large, covered and relatively visually inoffensive;

- waste is set out in containers so that the crew can pick them up quickly;

- health risk to the collectors and odour on the streets are minimised;

- waste is relatively inaccessible to the waste pickers.

Garbage Truck

Garbage truck or dustcart refers to a truck specially designed to collect municipal solid waste and haul the collected waste to a solid waste treatment facility such as a landfill. Other common names for this type of truck include trash truck in the United States, and rubbish truck, bin wagon, dustbin lorry, bin lorry or bin van elsewhere. Technical names include waste collection vehicle and refuse collection vehicle. These trucks are a common sight in most urban areas.

A Scania front loader

Major U.S. manufacturers of garbage trucks include Mack and Autocar Trucks. Major manufacturers of garbage truck bodies (not the truck itself) include McNeilus, and Heil.

History

Thornycroft Steam Dust-Cart of 1897 with tipper body

Wagons and other means had been used for centuries to haul away solid waste. Among the first self-propelled garbage trucks were those ordered by Chiswick District Council from the Thornycroft Steam Wagon and Carriage Company in 1897 described as a steam motor tip-car, a new design of body specific for "the collection of dust and house refuse".

The 1920s saw the first open-topped trucks being used, but due to foul odors and waste falling from the back, covered vehicles soon became more common. These covered trucks were first introduced in more densely populated Europe and then in North America, but were soon used worldwide.

The main difficulty was that the waste collectors needed to lift the waste to shoulder height. The first technique developed in the late 20s to solve this problem was to build round compartments with massive corkscrews that would lift the load and bring it away from the rear. A more efficient model was the development of the hopper in 1929. It used a cable system that could pull waste into the truck.

In 1937, George Dempster invented the *Dempster-Dumpster* system in which wheeled waste containers were mechanically tipped into the truck. His containers were known as Dumpsters, which led to the word dumpster entering the language.

In 1938, the Garwood Load Packer revolutionized the industry when the notion of including a compactor in the truck was implemented. The first primitive compactor could double a truck's capacity. This was made possible by use of a hydraulic press which compacted the contents of the truck periodically.

RS-3 Lightning Rear Steer truck

1955 saw the Dempster Dumpmaster the first front loader introduced, however they didn't become common until the 1970s. The 1970s also saw the introduction of smaller dumpsters, often known as wheelie bins which were also emptied mechanically. Since that time there has been little dramatic change, although there have been various improvements to the compaction mechanisms in order to improve payload. In the mid-1970s Petersen Industries introduced the first grapple truck for municipal waste collection.

In 1997, Lee Rathbun introduced the *Lightning Rear Steer System*. This system includes an elevated, rear-facing cab for both driving the truck and operating the loader. This configuration allows the operator to follow behind haul trucks and load continuously.

Types of Waste Collection Vehicle

A standard Waste Management Inc. front-loading garbage truck in San Jose, California

Front Loaders

Front loaders generally service commercial and industrial businesses using large waste containers with lids known as Dumpsters in the US. The truck is equipped with powered forks on the front which the driver carefully aligns with sleeves on the waste container using a joystick or a set of levers. The waste container is then lifted over the truck. Once it gets to the top the container is then flipped upside down and the waste or recyclable material is emptied into the vehicle's hopper. Once the waste is dumped, it is compacted by a hydraulically powered moving wall that oscillates backwards and forwards to push the waste to the rear of the vehicle. Most of the newer packing trucks have "pack-on-the-go hydraulics" which lets the driver pack loads while driving, allowing faster route times. When the body is full, the compaction wall moves all the way to the rear of the body, ejecting it via an open tailgate. There is also a system called the Curotto Can which is an attachment for a front loader that has an automated arm that functions as an automated side loader that allows the driver to dump carts.

14.5 m³ rear load container serviced in Copenhagen

Garbagemen loading garbage by hand in Japan, 2013

Rear Loaders

Rear loaders have an opening at the rear that a waste collector can throw waste bags or empty the contents of bins into. Often in many areas they have a lifting mechanism to automatically empty large carts without the operator having to lift the waste by hand.

Another popular system for the rear loader is a rear load container specially built to fit a groove in the truck. The truck will have a chain or cable system for upending the container. The waste will then slide into the hopper of the truck.

The modern rear loader usually compacts the waste using a hydraulically powered mechanism that employs a moving plate or shovel to scoop the waste out from the loading hopper and compress it against a moving wall. In most compactor designs, the plate has a pointed edge (hence giving it the industry standard name *packer blade*) which is designed to apply point pressure to the waste to break down bulky items in the hopper before being drawn into the main body of the truck.

Compactor designs, however, have been many and varied, however the two most popular in use today are the "sweep and slide" system (first pioneered on the Leach 2R Packmaster), where the packer blade pivots on a moving carriage which slides back and forth, and the "swing link" system (such as the Dempster Routechief) where the blade literally swings on a "pendulum"-style mechanism. The Heil Colectomatic used a combination of a lifting loading hopper and a pivoting sweeper blade to clear and compact waste in anticipation of the next load.

So-called "continuous" compactors were popular in the 1960s and 1970s. The German Shark design (later Rotopress) used a huge rotating drum, analogous to a cement mixer, in conjunction with a serrated auger to grind down and compact the garbage. SEMAT-Rey of France pioneered the rotating rake system (also used in the British Shelvoke and DrewryRevopak) to both mutilate waste and break down large items. Other systems used a large Archimedes' screw to draw in waste and mutilate it inside the body. A mixture of safety concerns, and higher fuel consumption has seen a decline in the popularity of continuously compacting garbage trucks. The Rotopress design remains popular due to its niche in being able to effectively deal with green waste for composting.

The wall will move towards the front of the vehicle as the pressure forces the hydraulic valves to open, or as the operator moves it with a manual control.

A unique rear-loading system involves a rear loader and a front-loading tractor (usually a Caterpillar front loader with a Tink Claw) for yard waste collection (and in some cities, garbage and recycling). The front loader picks up yard waste set in the street, and then loaded into the back of a rear loader. This system is used in several cities, including San Jose.

Side Loaders

Side loaders are loaded from the side, either manually, or with the assistance of a joystick-controlled robotic arm with a claw, used to automatically lift and tip wheeled bins into the truck's hopper.

Automated Side Loaders

Lift-equipped trucks are referred to as automated side loaders, or ASL's. Similar to a front-end loader, the waste is compacted by an oscillating packer plate at the front of the loading hopper which forces the waste through an aperture into the main body and is therefore compacted towards the rear of the truck. An Automated Side Loader only needs one operator, where a traditional rear load garbage truck may require two or three people, and has the additional advantage of reducing on the job injuries due to repetitive heavy lifting. Due to these advantages, ASL's have become more popular than traditional manual collection. Typically an Automated Side Loader uses standardized wheeled carts compatible with the truck's automated lift.

Semi-automated Side Loaders

An Automated side loader garbage truck in Canberra, Australia

Semi-automated side loaders use an automated mechanism to lift and dump manually aligned waste containers inside the main body of the truck. The primary difference of semi-automated side loaders is that they require more than one person, to operate the truck, and to manually bring and align containers to the loading hopper on the side of the truck.

Automated garbage collection in Aardenburg, Netherlands

As with front loaders, the compaction mechanism comprises a metal pusher plate in the collection hopper which oscillates backwards and forwards under hydraulic pressure, pushing the refuse through an aperture, thus compacting it against the material already loaded. On some ASL's there is also a "folding" crusher plate positioned above the opening in the hopper, that folds down to crush bulky items within reach of the metal pusher plate. Another compactor design is the "paddle packer" which uses a paddle that rotates from side to side, forcing refuse into the body of the truck.

Automated side loader in operation on an Autocar truck chassis.

Garbage collection by an automatic side loader during autumn in Kelowna, British Columbia, Canada

Pneumatic Collection

Volvo pneumatic collector used for "waste suction"

Pneumatic collection trucks have a crane with a tube and a mouthpiece that fits in a hole, usually hidden under a plate under the street. From here it will suck up waste from an underground installation. The system usually allows the driver to "pick up" the waste, even if the access is blocked by cars, snow or other barriers.

Grapple truck

Grapple Trucks

Grapple trucks enable the collection of bulk waste. A large percentage of items in the solid waste stream are too large or too heavy to be safely lifted by hand into traditional garbage trucks. These items (furniture, large appliances, branches, logs) are called bulky waste or "oversized." The preferred method for collecting these items is with a grapple truck. Grapple trucks have hydraulic knucklebooms, tipped with a clamshell bucket, and usually include a dump body or trailer.

Roll-offs

Roll-offs are characterized by a rectangular footprint, utilizing wheels to facilitate rolling the dumpster in place. The container is designed to be transported by special roll-off trucks. They are relatively efficient for bulk loads of waste.

Sand Cleaning Machine

Raking beach cleaner purifies sand

A tractor-pulled beach cleaner at Hietaniemi beach in Helsinki, Finland.

A sand cleaning machine, beach cleaner, or (colloquially) sandboni is a vehicle that drags a raking or sifting device over beach sand to remove rubbish and other foreign matter. They are manually self-pulled vehicles on tracks or wheels or pulled by quad-bike or tractor. Seaside cities use beach cleaning machines to combat the problems of litter left by beach patrons and other pollution washed up on their shores. A chief task in beach cleaning strategies is finding the best way to handle waste matter on the beaches, taking into consideration beach erosion and changing terrain. Beach cleaning machines work by collecting sand by way of a scoop or drag mechanism and then raking or sifting anything large enough to be considered foreign matter, including sticks, stones, litter and other items. Similar applications include lake beaches, sandfields for beach volleyball and kindergarten and playing field sandpits. The word "sandboni" is a back-formation referencing the ice-surfacing machine Zamboni.

Common Technologies

Raking technology can be used on dry or wet sand. When using this method, a rotating conveyor belt containing hundreds of tines combs through the sand and removes surface and buried debris while leaving the sand on the beach. Raking machines can remove materials ranging in size from small pebbles, shards of glass, and cigarette butts to larger debris, like seaweed and driftwood. By keeping the sand on the beach and only lifting the debris, raking machines can travel at high speeds.

Sifting technology is practiced on dry sand and soft surfaces. The sand and waste are collected via the pick-up blade of the vehicle onto a vibrating screening belt, which leaves the sand behind. The waste is gathered in a collecting tray which is often situated at the back of the vehicle. Because sand and waste are lifted onto the screening belt, sifters must allow time for the sand to sift through the screen and back onto the beach. The size of the materials removed is governed by the size of the holes in the installed screen.

Combined raking and sifting technology differs from pure sifters in that it uses rotating tines to scoop sand and debris onto a vibrating screen instead of relying simply on the pick-up blade. The tines' position can be adjusted to more effectively guide different-sized materials onto the screen. Once on the screen, combined raking and sifting machines use the same technology as normal sifters to remove unwanted debris from the sand.

Sand sifting by hand is used for smaller areas or sensitive habitat. Sand and debris is collected into a windrow or pile and manually shoveled onto screened sifting trays to separate the debris from the sand. While effective, it requires the movement of sand to the site of the tray, and then redistribution of the sand after sifting. A more efficient method is the use of a screened fork at the place where the debris is located. The effort to manually agitate the sand can become tiresome; however, a recent development of a battery-powered sand rake combines the spot cleaning effectiveness of manual screening with the ease of an auto-sifting hand tool.

Areas of operation

Sandboni at the gulf of Mexico

Beach volleyball field after rework of a sand cleaner

Sand cleaning machines are used all over the world to ensure the safety and happiness of beach-goers. By removing litter, unwanted seaweed, and other debris from the beach, municipalities and resorts are able to maintain their beaches with fewer invested hours.

In addition to their regular litter-removing uses, beach and sand cleaners have been used to clean up after natural disasters. For example:

In Galveston, Texas, low oxygen levels in the water resulted in thousands of dead fish washing ashore. Raking sand cleaners were then used to remove the rotting fish off the beach before they released excessive toxins into the air, sand, and water.

The Olympic Games 2008 saw the first remote-control Sandbonis for the beach volleyball fields in Beijing Chaoyang Park.

The cleanup after the Deepwater Horizon oil spill saw large applications of sand cleaners to the area. Similarly, the Rena oil spill in New Zealand also saw beach cleaners deployed in an effort to remove the affected sand.

Manufacturers

The major manufacturers of large beach-sand cleaning machines are considered to be H Barber & Sons, Cherrington, Beach Tech, Rockland and Tirrenia Srl.

There are many other manufacturers of sand cleaners being used for other purposes. For example, a smaller 4-wheel and halftrack sand cleaning machine is used for sandpits in Kindergarten and municipality playing fields and for beach volleyball. When environmental or spot-cleaning requires hand operations, an auto-sifting, lightweight, screened rake can be the best choice

Waste Sorting

Recycling bins in Singapore

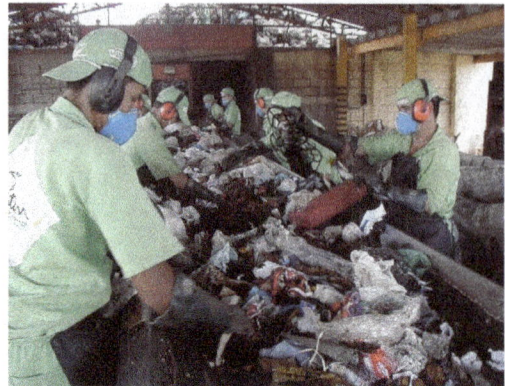

Manual waste sorting for recycling

Characteristic containers for recycling in Portovenere, Italy

Garbage containers in Fuchū, Tokyo, Japan

Waste sorting is the process by which waste is separated into different elements. Waste sorting can occur manually at the household and collected through curbside collection schemes, or automatically separated in materials recovery facilities or mechanical biological treatment systems. Hand sorting was the first method used in the history of waste sorting.

Waste Can also be Sorted in a Civic Amenity Site.

Waste segregation means dividing waste into dry and wet. Dry waste includes wood and related products, metals and glass. Wet waste, typically refers to organic waste usually generated by eating establishments and are heavy in weight due to dampness. Waste can also be segregated on basis of biodegradable or non-biodegradable waste.

Landfills are an increasingly pressing problem. Less and less land is available to deposit refuse, but the volume of waste is growing all time. As a result, segregating waste is not just of environmental importance, but of economic concern, too.

Methods

Waste is collected at its source in each area and separated. The way that waste is sorted must reflect local disposal systems. The following categories are common:

- Paper

- Cardboard (including packaging for return to suppliers)

- Glass (clear, tinted – no light bulbs or window panes, which belong with residual waste)

- Plastics

- Textiles

- Wood, leather, rubber

- Scrap metal

- Compost

- Special/hazardous waste

- Residual waste

Organic waste can also be segregated for disposal:

- Leftover food which has had any contact with meat can be collected separately to prevent the spread of bacteria.

  o Meat and bone can be retrieved by bodies responsible for animal waste

  o If other leftovers are sent, for example, to local farmers, they can be sterilised before being fed to the animals

- Peel and scrapings from fruit and vegetables can be composted along with other degradable matter. Other waste can be included for composting, too, such as cut flowers, corks, coffee grindings, rotting fruit, tea bags, egg- and nutshells, paper towels etc.

Chip pan oil (fryer oil), used fats, vegetable oil and the content of fat filters can be collected by companies able to re-use them. Local authority waste departments can provide relevant addresses. This can be achieved by providing recycling bins.

By Country

In Germany, regulations exist that provide mandatory quotas for the waste sorting of packaging waste and recyclable materials such as glass bottles.

In Denpasar, Bali, Indonesia, a pilot project using an automated collecting machine of plastic bottles or aluminium cans with voucher reward has been implemented in a market.

Transfer Station

Transfer station is a centralised facility, where waste is unloaded from smaller collection vehicles and re-loaded into large vehicles for transport to a disposal or processing site. This transfer of waste is frequently accompanied by removal, separation or handling of waste. In areas, where wastes are not already dense, they maybe compacted at a transfer station. The technical limitations of smaller collection vehicles and thelow hauling cost of solid waste, using larger vehicles, make a transfer station viable. Also, the use of transfer station proves reasonable, when there is a need for vehicles servicing a collection route to travel shorter distances, unload and return quickly to their primary task of collecting the waste.

Limitations in hauling solid wastes are the main factors to be considered, while evaluating the use of transfer stations. These include the additional capital costs of purchasing trailers, building transfer stations and the extra time, labour and energy required for transferring wastes from collection truck to transfer trailer.

Consider also the following factors that affect the selection of a transfer station:

- Types of waste received.

- Processes required in recovering material from wastes.

- Required capacity and amount of waste storage desired.

- Types of collection vehicles using the facility.

- Types of transfer vehicles that can be accommodated at the disposal facilities.

- Site topography and access.

The main problem in the establishment of a transfer station, however, is securing a suitable site. Stored solid wastes and recyclable materials, if not properly handled, will attract flies and other insect vectors. Odours from the transferred solid wastes will also be a nuisance, if not properly controlled. In addition, the traffic and noise due to small and large collection vehicles, collectors, drivers, etc., invite the resentment of the communities living in the vicinity of transfer stations (EPA, 1995).

Types

Depending on the size, transfer stations can be either of the following two types:

(i) Small to medium transfer stations: These are direct-discharge stations that provide no intermediate waste storage area. The capacities are generally small (less than 100 tonnes/day) and medium (100 to 500 tonnes/day). Depending on weather, site aesthetics and environmental concerns, transfer operations of this size may be located either indoor or outdoor. More complex small transfer stations are usually attended during hours of operation and may include some simple waste and materials processing facilities. For example, it includes a recyclable material separation and processing centre. The required overall stationcapacity (i.e., the number and size of containers) depends on the size and population density of the area served and the frequency of collection.

(ii) Large transfer stations: These are designed for heavy commercial use by private and municipal collection vehicles. The typical operational procedure for a larger station is as follows:

- when collection vehicles arrive at the site, they are checked in for billing, weighed and directed to the appropriate dumping area;

- collection vehicles travel to the dumping area and empty the wastes into a waiting trailer, a pit or a platform;

- after unloading, the collection vehicle leaves the site, and there is no need to weigh the departing vehicle, if its weight (empty) is known;

- transfer vehicles are weighed either during or after loading. If weighed during loading, trailers can be more consistently loaded to just under maximum legal weights and this maximises payloads and minimises weight violations.

Designs for Larger Transfer Operations

Several different designs for larger transfer operations are common, depending on the transfer distance and vehicle type. Most designs, however, fall into one of the following three categories:

(i) Direct-discharge non-compaction station: In these stations, waste is dumped directly from collection vehicle into waiting transfer trailers and is generally designed with two main operating floors. In the transfer operation, wastes are dumped directly from collection vehicles (on the top floor) through a hopper and into open top trailers on the lower floor. The trailers are often positioned on scales so that dumping can be stopped when the maximum payload is reached. A stationary crane with a bucket is often used to distribute the waste in the trailer. After loading, a cover or tarpaulin is placed over the trailer top. However, some provision for waste storage during peak time or system interruptions should be developed. Because of the use of little hydraulic equipment, a shutdown is unlikely and this station minimises handling of waste.

(ii) Platform/pit non-compaction station: In this arrangement, the collection vehicles dump their wastes onto a platform or into a pit using waste handling equipment, where wastes can be temporarily stored, and if desired, picked through for recyclables or unacceptable materials. The waste is then pushed into open-top trailers, usually by front-end loaders. Like direct discharge stations, platform stations have two levels. If a pit is used, however, the station has three levels. A major advantage of these stations is that they provide temporary storage, which allows peak inflow of wastes to be levelled out over a longer period. Construction costs for this type of facility are usually higher because of the increased floor space. This station provides convenient and efficient storage area and due to simplicity of operation and equipment, the potential for station shutdown is less.

(iii) Compaction station: In this type of station, the mechanical equipment is used to increase the density of wastes before they are transferred. The most common type of compaction station uses a hydraulically powered compactor to compress wastes. Wastes are fed into the compactor through a chute, either directly from collection trucks or after intermediate use of a pit. The hydraulic ram of the compactor pushes waste into the transfer trailer, which is usually mechanically linked to the compactor (EPA, 1995). Compaction stations are used when:

- wastes must be baled for shipment;

- open-top trailers cannot be used because of size restrictions;

- site topography or layout does not accommodate a multi-level building.

The main disadvantage of a compaction facility is that the facility's ability to process wastes is directly dependent on the operative-ness of the compactor. Selection of a quality compactor, regular maintenance of the equipment, easy availability of spare parts and prompt availability of the service personnel are essential for the station's reliable operation.

Capacity

A transfer station should have enough capacity to manage and handle the wastes at the facility throughout its operating life. While selecting the design capacity of a transfer station, we must, therefore, consider trade-offs between the capital costs associated with the station and equipment and the operational costs. Designers should also plan adequate space for waste storage and, if necessary, waste processing. Transfer stations are usually designed to have 1.5– 2 days of storage capacity. The collection vehicle unloading area is usually the waste storage area and sometimes a waste sorting area. When planning the unloading area, designers should allow adequate space for vehicle and equipment manoeuvring. To minimise the space required, the facility should be designed such that the collection vehicle backs into the unloading position. Adequate space should also be available for offices, employee facilities, and other facility-related activities (EPA, 1995). Factors that should be considered in determining the appropriate capacity of a transfer facility include:

- capacity of collection vehicles using the facility;

- desired number of days of storage space on tipping floor; time required to unload collection vehicles;

- number of vehicles that will use the station and their expected days and hours of arrival;

- waste sorting or processing to be accomplished at the facility; transfer trailer capacity;

- hours of station operation;

- availability of transfer trailers waiting for loading;

- time required, if necessary, to attach and disconnect trailers from tractors or compactors.

Transfer station capacity can be determined using the following formulae:

(i) Pit stations: Based on the rate at which wastes can be unloaded from collection vehicles:

$$C = P_c \times (L/W) \times (60 \times H_w/T_c) \times F$$

Based on rate at which transfer trailers are loaded:

$$C = (P_t \times N \times 60 \times H_t)/(T_t + B)$$

(ii) Direct dump stations:

$$C = (Nn \times Pt \times F \times 60 \times Hw)/[((Pt/Pc) \times (W/Ln)) \times Tc + B]$$

(iii) Hopper compaction stations:

$$C = (N_n \times P_t \times F \times 60 \times H_w)/[(P_t/P_c \times T_c) + B]$$

(iv) Push pit compaction station:

$$C = (N_p \times P_t \times F \times 60 \times H_w)/[(P_t/P_c \times W/L_p \times T_c) + B_c + B]$$

Where:

C = Station capacity (tonnes/day) ; P_c= Collection vehicle payload (tonnes) ; L = Total length of dumping space (feet) ; H_w = Hours per day that waste is delivered; T_c = Time (in minutes) to unload each collection vehicle; F = Peaking factor (ratio of the number of collection vehicles received during an average 30-minute period to the number received during a peak 30-minute period) ; L_p = Length of push pit (feet) ; N_p = Number of push pits; B_c = Total cycle time for clearing each push pit and compacting waste into trailer; P_t = Transfer trailer payload (tonnes) ; N = Number of transfer trailers loading simultaneously; H_t = Hours per day used to load trailers (minutes) ; B = Time to remove and replace each loaded trailer (minutes) ; T_t = Time to load each transfer trailer (minutes) ; N_n = Number of hoppers; L_n = Length of each hopper (feet).

These formulae are useful in estimating the capacity of various types of transfer stations (EPA, 1995) and should be adapted, as necessary, for specific applications.

Viability

Transfer stations offer benefits such as lower collection costs (because crews waste less time travelling to the site), reduced fuel and maintenance costs for collection vehicles, increased flexibility in selection of disposalfacilities, opportunity to recover recyclables or compostables at the transfer site and the opportunity to shred or scoop wastes prior to disposal. These benefits must be weighed against the costs to develop and operate the facility.

The classical approach to arrive at the economic viability of operating a transfer station, is to add the unit cost of the transfer station to the cost of hauling using large vehicles, and to compare this cost with the cost of hauling directly to the disposal site using the smaller vehicles that service the collection area. The cost of hauling using small vehicles is the sum of the depreciation cost of the vehicle, driver's salary, salary of the collection crew (if they are on standby waiting for the vehicle to return to the collection area) and fuel cost. The transfer station cost is the sum of the transfer station's depreciation cost and the operating and maintenance costs divided by the capacity of the station. The cost of using the large vehicle is the sum of the vehicle depreciation, fuel cost and driver's salary.

The cost-effectiveness of a transfer station depends on the distance of disposal site from thegeneration area, andadistance of 10 – 15 km is usually the minimum cost-effective distance (Phelps, et al., 1995). The distance between the disposal site and collection area is one of the principal variables in deciding whether to use a transfer station or haul the solid wastes directly from the collection area to the disposal site. Figure illustrates the economic analysis involving the effect of the hauling distance on the collection cost:

Cost Analysis to Determine Viability of Transfer Station

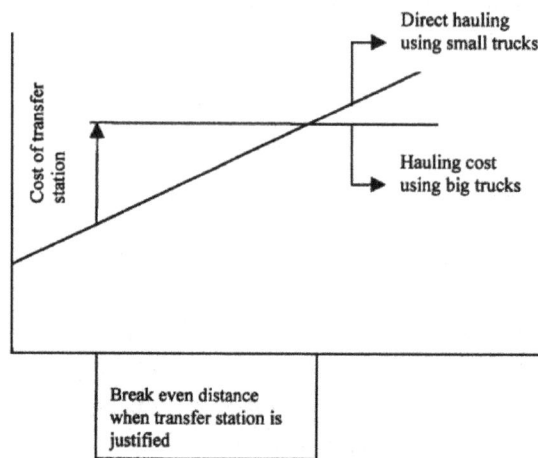

Now, let us consider first the case in which the transfer station is located direc tly along the hauling route between the disposal site and the collection area. Let the unit cost of hauling using a small vehicle be Rs. A/m3 km. The cost of operation, maintenance, depreciation, loading and unloading at the transfer station be Rs. B/m3 and the cost of hauling using large vehicles beRs.C/m3 km.Ifthe distance between the collection area and the transfer station is X km and the distance between the transfer station and the disposal site is Y km, then the distance between the collection area and the disposal site is X + Y km. Then, the total cost of hauling the solid wastes from the collection area to the disposal site using a transfer station is:

$$T = 2AX + B + 2CY$$

The factor 2 is added to account for the round trip, which effectively doubles the distance travelled. The total cost of hauling without the transfer station is:

$$T_1 = 2A(X + Y)$$

The transfer station is justified, when:

$$T < T_1$$

That is, the hauling cost using a transfer station is lower than the direct hauling costs between the collection area and the disposal site. Substituting the values of T and T_1 yields:

$$2AX + B + 2CY < 2AX + 2AY$$

or

$$Y > B/(2A - 2C)$$

Note that X cancels out. The distance between the potential transfer station site and the disposal site is the variable to consider. The distance between the collection area and the disposal site is important in deciding the utilisation of a transfer station, if X is equal to zero, in which case the transfer station is located right at the centroid of the collection area. Under normal conditions, the centroid of the collection area has a high land value, and it would be impractical to locate a solid waste transfer station in this area. Figure shows the effect of the distance between the potential transfer station site and the disposal site on the hauling cost.

Consider a general case in which the transfer station is located away from the hauling route between the collection area and the disposal site. Let Z be the additional distance travelled by the vehicles. The cost T, when using a transfer station, is then equal to:

$$T = B + 2AX + 2AZ + 2CY + 2CZ$$

The cost of direct hauling from the collection area to the disposal site remains the same as previously defined.The use of a transfer station is justified, if:

$$B + 2AX + 2AZ + 2CY + 2AZ < 2AX + 2AY$$

or

$$Y > (B + 2CZ + 2AZ)/(2A - 2C)$$

Record Keeping, Control, Inventory and Monitoring

For effective waste collection and, indeed, SWM, we must maintain records on the quantities of wastes collected and their variation within a week, month and year, as well as on established long-term trends in solid waste generation ratesand composition, sources of wastes and the personnel collecting them. Long- term trends in solid waste generation rates and composition form the basis for planning, especially in budgeting for future vehicle requirements, allocating the collection vehicles and crew, building transfer stations, acquiring strategic lands and determining disposal options. Table contains an illustration of a checklist of factors that affect the waste collection system:

Checklist of Variables Affecting Collection System

Components	Factors to Consider
Crew size	labour cost distance between containers size and types of containers loading accessories available in the truck collection vehicle used
Container type	solid wastes generation rate density of waste generation street width traffic vol ume collection crew configuration standard of li ving

Collection accessory	labour cost protection of worker's health
Vehicle size/type	streetwidth, trafficvolumesolidwastegenerationratescrewsize viabilityofatransferstation
Collection route	street width, traffic volume solid waste generation rates crew size viability of a transfer station street width, traffic volume direction of traffic flow solid waste generation rates spatial distribution of wastes local topography
Transfer station	distance between disposal site and collection area hauling cost for small and large trucks cost of transferring the solid wastes from small to large trucks

Records of personnel and quantities of wastes collected are, when maintained, useful in determining the efficiency of the personnel and in correlating waste quantities with conditions in the service area. A time keeping system at the transfer or disposal site is a key element in improving the efficiency of collection system and planning an upgraded system. The timekeeping system determines if the crewwere taking long rest periods, spending time salvaging or carrying out unauthorised activities. The performance of a particular crew in terms of the quantity of solid wastes collected per day could be compared with that of another collection crew working under similar conditions.

The composition of solid wastes should be measured at least once a year for major districts and possibly once every two years in residential areas with stagnant growth rates and development. Changes in composition affect the collection equipment and configuration of the collection system is important in designing the disposal system. Changes in an energy source (such as a shift to gas or electricity from wood or charcoal for cooking and heating), reduces the ash content of wastes, making the solid waste lighter, in which case, larger containers could be used. The same line of analysis holds true in specifying the collection vehicles. Comparison of the routes taken by various crew serving a particular area helps to identify the best hauling route. Although this route may be longer, it could be more economical in terms of hauling time. However, note that the best route often changes with the season.

All these decisions should be based on reliable data, without which the waste collection system will inevitably be ineffective. Proper interpretation of monitoring data allows the authority to adapt the proposed system to actual conditions. In some instances, it also allows management to identify areas, where the design is not realistic.

Implementing Collection and Transfer System

Implementing of collection and transfer system involves the following activities, which are important for success of the plan (EPA, 1995):

(i) Finalising and implementing the system management plan: For proper implementation of collection and transfer system, it is necessary to have clear organisational structures and management plans. The organisational structure shouldbe simple, with a minimumofadministrative andmanagement layers between collection crews and top management. All workers in the department should clearly understand the department's mission and their roles. Through training, incentives and reinforcement by management, workers should be encouraged to be customer-oriented and team contributors. Feedback mechanisms must be introduced to help the crew review their performance and help managers monitoring the performance of crews, equipment, etc. It is also important to periodically review the management plans and structures, as implementation of collection services continues.

(ii) Purchasing and managing equipment: For purchasing equipment, most municipalities issue bid specifications. Detailed specifications include exact requirements for equipment sizes and capacities, power ratings, etc. Performance specifications often request that equipment be equivalent to certain available models and meet standards for capacity, speed, etc. Municipalities may either perform equipment maintenance themselves, contract with a local garage, or in some cases, contract with the vehicle vendor at the time of purchase. As part of the preventive maintenance programme, the collection crew should check the vehicle chassis, tyres and body daily and report any problems to maintenance managers. In addition, each vehicle should have an individual maintenance record that includes the following items:

- preventive maintenance schedule;

- current list of specific engine;

- a description of repairs and a list containing information on the repair date, mechanic, cost, type and manufacturer of repair parts and the length of time the truck was out of service, for each maintenance event.

(iii) Hiring and training personnel: As in all organisations, good personnel management is essential to an efficient, high-quality waste collection system. Authorities responsible for SWM should, therefore, strive to hire and keep well-qualified personnel. The recruitment programme shouldassess applicants' abilities to perform the types of physical labour required for the collection, equipment and methods used. To retain employees, management should provide a safe working environment that emphasises career advancement, participatory problem solving and worker incentives. Worker incentives should be developed to recognise and reward outstanding performance by employees. Ways to accomplish motivation include merit-based compensation, awards programme and a work structure. Feedback on employee performance should be regular and frequent.

Safety is especially important because waste collection employees encounter many hazards during each workday. As a result of poor safety records, insurance costs for many collection services are high. To minimise injuries, haulers should have an ongoing safety programme. This programme should outline safety procedures and ensure that all personnel are properly trained on safety issues. Haulers should develop an employee- training programme that helps employees improve and broaden the range of their job-related skills. Education should address such subjects as driving skills, first aid, safe lifting methods, identification of household hazardous wastes, avoidance of substance abuse and stress management.

(iv) Providing public information: Maintaining good communication with the public is important to a well-run collection system. Residents can greatly influence the performance of the collection system by co-operating in separation requirements, and by keeping undesirable materials from entering the collected waste stream. Commonly used methods of communicating information include brochures, articles in community newsletters, newspaper articles, announcements, and advertisements on radio and television, information attachments to utility bills (either printed or given separately) and school handouts. Communication materials should be used to help residents understand the community waste management challenges and the progress in meeting them. Residents should also be kept informed about issues such as the availability and costs of landfillcapacity so that they develop an understanding of the issues and a desire to help meet their waste management needs.

(v) Monitoring system cost and performance: Collection and transfer facilities should develop and maintain an effective system for cost and performance reporting. Each collection crew should complete a daily report containing the following information:

- Total quantity hauled.

- Total distance and travel times to and from the disposal site.

- Amounts delivered to each disposal, transfer, or processing facility.

- Waiting time at sites.

- Number of loads hauled.

- Vehicle or operational problems needing attention.

Collected data should be used to forecast workloads, truck costs, identify changes in the generation of wastes and recyclables, trace the origin of problematic waste materials and evaluate crew performance. Just as the goals of a collection programme set its overall directions, a monitoring system provides the short-term feedback necessary to identify the corrections needed to achieve those goals.

In brief guidelines for planning waste collection and transport are given below:

- Analyse the quantum of waste generated with composition.

- Capacity building of town municipalities with appropriate infrastructure and the knowledge of existing laws or regulations on waste collection, transport and safe disposal.

- Designate a para-state agency to oversee waste collection, transport and disposal to avoid confusion among para-state government agencies.

- Determine geographic scope of collection and transport services. Determine funding, equipment and labour needs.

- Determine the type and amount of waste to be processed

- Implement decentralised waste treatment through proven local techniques.

- Deploy GPS (Global Positioning System) based trucks for waste collection and transport to minimise pilferages.

- Adopt spatial information system for the management.

- Consider a transfer station that serves as a central location for activities to sort and recover waste.

- Implement decentralised waste management including all stakeholders with active participation of the public.

References

- "The State of New Zealand's Environment". Ministry for the Environment (New Zealand). 1997. Retrieved 2008-03-27

- Ministry for the Environment (December 2007). Environment New Zealand 2007. Ministry for the Environment (New Zealand). ISBN 978-0-478-30192-2

- "Memorandum submitted by Essex Friends of the Earth". www.publications.parliament.uk. Retrieved 19 January 2016

- Motor-Cars for Dust Collection", The Automotor and Horseless Carriage Journal, February 1897, p192

- Geroux, Zachary; Voytko, Eric. "The Ever Expanding History of the Front Load Refuse Truck". Retrieved 10 September 2014

Solid Waste Disposal: Practices and Issues

Waste that is collected needs to be properly disposed, as it can otherwise create health hazards and environmental issues. Some of the ways of waste disposal are sanitary landfill, composting, incineration, gasification, pyrolysis and refuse-derived fuel. This chapter is an overview of the subject matter incorporating all the major aspects of solid waste disposal.

Disposal Options and Selection Criteria

The most common disposal option practised currently in many countries is either uncontrolled dumping or dumping with moderate control. The environmental costs of uncontrolled dumping include breeding of disease causing vectors (e.g., flies, mosquitoes and rodents), pollution, odour and smoke.

Disposal Options

In this Section, the options available for waste disposal is discussed, and in that respect, we will consider the following:

(i) Uncontrolled dumping or non-engineered disposal: As mentioned, this is the most common method being practised in many parts of the world, and India is no exception. In this method, wastes are dumped at a designated site without any environmental control. They tend to remain there for a long period of time, pose health risks and cause environmental degradation. Due to the adverse health and environmental impact associated with it, the non-engineered disposal is not considered a viable and safe option.

(ii) Sanitary landfill: Unlike the non-engineered disposal, sanitary landfill is a fully engineered disposal option in that the selected location or wasteland is carefully engineered in advance before it is pressed into service. Operators of sanitary landfills can minimise the effects of leachate (i.e., polluted water which flows from a landfill) and gas production through proper site selection, preparation and management. This particular option of waste disposal is suitable when the land is available at an affordable price, and adequate workforce and technical resources are available to operate and manage the site.

(iii) Composting: This is a biological process of decomposition in which organisms, under controlled conditions of ventilation, temperature and moisture, convert the organic portion of solid waste into humus-like material. If this process is carried out effectively, what we get as the final product is a stable, odour-free soil conditioner. Generally, the option of composting is considered, when a considerable amount of biodegradable waste is available in the waste stream and there is use or market for composts. Composting can be either centralised orsmall-scale. Centralised composting plants are possible, if adequate skilled workforce and equipments are available. And, small-scale composting practices can be effective at house-hold level, but this needs public awareness.

(iv) Incineration: This refers to the controlled burning of wastes, at a high temperature (roughly 1200 – 1500 C), which sterilises and stabilises the waste in addition to reducing its volume. In the process, most of the combustible materials(i.e.,self-sustainingcombus-tible matter, whichsavestheenergyneeded tomaintain the combustion) such as paper or plastics get converted into carbon dioxide and ash. Incineration may be used as a disposal option, when land filling is not possible and the waste composition is highly combustible. An appropriate technology, infrastructure and skilled workforce are required to operate and maintain the plant.

(v) Gasification: This is the partial combustion of carbonaceous material (through combus-tion) at high temperature (roughly 1000 C) forming a gas, comprising mainly carbon diox-ide, carbon monoxide, nitrogen, hydrogen, water vapour and methane, which can be used as fuel.

(vi) Refuse-derived fuel (RDF): This is the combustible part of raw waste, separated for burning as fuel. Various physical processes such as screening, size reduction, magnetic separation, etc., are used to separate the combustibles.

(vii) Pyrolysis: This is the thermal degradation of carbonaceous material to gaseous, liquid and solid fraction in the absence of oxygen. This occurs at a temperature between 200 and 900 C.The product of pyrolysis is a gas ofrelatively high calorific value of 20,000 joules per gram with oils, tars and solid burned residue (Ali, et al 1999).

Relative Merits of Some Options

Having touched upon several disposal options, let us now present the merits and demerits of some of them in Table:

Relative Merits of Disposal Options

Disposal Option / Sustainability Indicator	Non- engineered Disposal	Sanitary Landfill	Composting	Incineration
Volume reduction	×	×	×	√
Expensive	×	√	√	√
Long term maintenance	√	√	×	×
By product recovery	×	√	√	√
Adaptability to all wastes	√	√	×	×
Adverse environmental effect	√	√	×	√

Selection Criteria

With the help of proper frameworks and sub-frameworks, we can assess the effectiveness of each of the waste disposal options. While a framework represents an aid to decision-making and helps to ensure the key issues are considered, a sub-framework explains how and why the necessary information should be obtained (Ali, et al 1999). A framework contains a list of issues and questions pertaining to the technical, institutional, financial, social and environmental features of a waste disposal system to assess the capacity of a disposal option to meet the requirements. For example, an appraisal of waste disposal option must include the following:

(i)   Technical: This feature, involving efficient and effective operation of the technology being used, evaluates the following components of a SWM system:

  • composition of wastes, e.g., type, characteristics and quantity.

  • existing practices, e.g., collection, transport, and recycling process.

  • siting, e.g., location of disposal site, engineering material, etc.

  • technology, e.g., operation, maintenance, technical support, etc.

  • impact, e.g., anticipated by-product, requirement for their treatment and disposal, etc

(ii) Institutional: This involves the ability and willingness of responsible agencies to operate and manage the system by evaluating the following:

  • structures,roles and responsibilities, e.g., currentinstitutional frameworks.

  • operational capacity, e.g., municipal capacities, local experience and staff training.

  • incentives, e.g.,management improvementandwastedisposal practices.

  • innovation and partnership.

(iii) Financial: This assesses the ability to finance the implementation, operation and maintenance of the system by evaluating the following:

  • financing and cost recovery, e.g., willingness to raise finance for waste management.

  • current revenue and expenditure on waste management.

  • potential need for external finance for capital cost.

(iv) Social: This helps in avoiding adverse social impact by evaluating the following:

  • waste picking, which has an impact on livelihood and access to waste pickers.

  • health and income implication.

  • public opinions on the existing and proposed system.

(v)   Environmental: This means setting up an environment friendly disposal system by evaluating the following:

- initial environmental risks, i.e., impact of existing and proposed disposal option.

- long-term environmental risks, i.e., long-term implication (future impacts).

Landfill

A landfill in Poland

A landfill site (also known as a tip, dump, rubbish dump, garbage dump or dumping ground and historically as a midden) is a site for the disposal of waste materials by burial and the oldest form of waste treatment (although the burial part is modern; historically, refuse was just left in piles or thrown into pits). Historically, landfills have been the most common method of organized waste disposal and remain so in many places around the world.

Some landfills are also used for waste management purposes, such as the temporary storage, consolidation and transfer, or processing of waste material (sorting, treatment, or recycling).

A landfill also may refer to ground that has been filled in with rocks instead of waste materials, so that it can be used for a specific purpose, such as for building houses. Unless they are stabilized, these areas may experience severe shaking or soil liquefaction of the ground during a large earthquake.

Operations

Typically, operators of well-run landfills for non-hazardous waste meet predefined specifications by applying techniques to:

1. confine waste to as small an area as possible

2. compact waste to reduce volume

3. cover waste (usually daily) with layers of soil

One of several landfills used by Dryden, Ontario, Canada.

During landfill operations a scale or weighbridge may weigh waste-collection vehicles on arrival and personnel may inspect loads for wastes that do not accord with the landfill's waste-acceptance criteria. Afterward, the waste-collection vehicles use the existing road network on their way to the tipping face or working front, where they unload their contents. After loads are deposited, compactors or bulldozers can spread and compact the waste on the working face. Before leaving the landfill boundaries, the waste collection vehicles may pass through a wheel-cleaning facility. If necessary, they return to the weighbridge for re-weighing without their load. The weighing process can assemble statistics on the daily incoming waste-tonnage, which databases can retain for record keeping. In addition to trucks, some landfills may have equipment to handle railroad containers. The use of "rail-haul" permits landfills to be located at more remote sites, without the problems associated with many truck trips.

Typically, in the working face, the compacted waste is covered with soil or alternative materials daily. Alternative waste-cover materials include chipped wood or other "green waste", several sprayed-on foam products, chemically "fixed" bio-solids, and temporary blankets. Blankets can be lifted into place at night and then removed the following day prior to waste placement. The space that is occupied daily by the compacted waste and the cover material is called a daily cell. Waste compaction is critical to extending the life of the landfill. Factors such as waste compressibility, waste-layer thickness and the number of passes of the compactor over the waste affect the waste densities.

Advantages

Landfills are often the most cost-efficient way to dispose of waste, especially in countries like the United States with large open spaces. While resource recovery and incineration both require extensive investments in infrastructure, and material recovery also requires extensive manpower to maintain, landfills have fewer fixed—or ongoing—costs, allowing them to compete favorably. In addition, landfill gas can be upgraded to natural gas—landfill gas utilization—which is a potential revenue stream. Another advantage is having a specific location for disposal that can be monitored, where waste can be processed to remove all recyclable materials before tipping.

Social and Environmental Impact

Landfill operation in Hawaii. Note that the area being filled is a single, well-defined "cell" and that a protective landfill liner is in place (exposed on the left) to prevent contamination by leachates migrating downward through the underlying geological formation.

Landfills have the potential to cause a number of issues. Infrastructure disruption, such as damage to access roads by heavy vehicles, may occur. Pollution of local roads and water courses from wheels on vehicles when they leave the landfill can be significant and can be mitigated by wheel washing systems. Pollution of the local environment, such as contamination of groundwater or aquifers or soil contamination may occur, as well.

Leachate

Extensive efforts are made to capture and treat leachate from landfills before it reaches groundwater aquifers, but engineered liners always have a lifespan, though it may be 100 years or more. Eventually, every landfill liner will leak, allowing the leachate to contaminate the groundwater. Installation of composite liners with flexible membrane and soil barrier is enforced by the EPA to ensure that leachate is withheld.

Dangerous Gases

Rotting food and other decaying organic waste allows methane and carbon dioxide to seep out of the ground and up into the air. Methane is a potent greenhouse gas, and can itself be a danger because it is flammable and potentially explosive. In properly managed landfills, gas is collected and utilized. This could range from simple flaring to landfill gas utilization. Carbon dioxide is the most widely produced greenhouse gas. It traps heat in the atmosphere, contributing to climate change.

Infections

Poorly run landfills may become nuisances because of vectors such as rats and flies which can cause infectious diseases. The occurrence of such vectors can be mitigated through the use of daily cover.

Other potential issues include wildlife disruption, dust, odor, noise pollution, and reduced local property values.

Landfill Gas

Gases are produced in landfills due to the anaerobic digestion by microbes. In a properly managed landfill this gas is collected and used. Its uses range from simple flaring to the landfill gas utilization and generation of electricity. Landfill gas monitoring alerts workers to the presence of a build-up of gases to a harmful level. In some countries, landfill gas recovery is extensive; in the United States, for example, more than 850 landfills have active landfill gas recovery systems.

Regional Practice

A landfill in Perth, Western Australia

South East New Territories Landfill, Hong Kong

Canada

Landfills in Canada are regulated by provincial environmental agencies and environmental protection acts (EPA). Older facilities tend to fall under current standards and are monitored for leaching. Some former locations have been converted to parkland.

European Union

In the European Union, individual states are obliged to enact legislation to comply with the requirements and obligations of the European Landfill Directive. In the UK this is the Waste Implementation Programme.

United Kingdom

Landfilling practices in the UK have had to change in recent years to meet the challenges of the European Landfill Directive. The UK now imposes landfill tax upon biodegradable waste which is put into landfills. In addition to this the Landfill Allowance Trading Scheme has been established for local authorities to trade landfill quotas in England. A different system operates in Wales where authorities are not able to 'trade' between themselves, but have allowances known as the Landfill Allowance Scheme.

United States

U.S. landfills are regulated by each state's environmental agency, which establishes minimum guidelines; however, none of these standards may fall below those set by the United States Environmental Protection Agency (EPA).

Permitting a landfill generally takes between 5 and 7 years, costs millions of dollars and requires rigorous siting, engineering and environmental studies and demonstrations to ensure local environmental and safety concerns are satisfied.

Microbial Topics

The status of a landfill's microbial community may determine its digestive efficiency.

Bacteria that digest plastic have been found in landfills.

Reclaiming Materials

Landfills can be regarded as a viable and abundant source of materials and energy. In the developing world, waste pickers often scavenge for still-usable materials. In a commercial context, landfill sites have also been discovered by companies, and many have begun harvesting materials and energy . Well known examples are gas recovery facilities. Other commercial facilities include waste incinerators which have built-in material recovery. This material recovery is possible through the use of filters (electro filter, active carbon and potassium filter, quench, HCl-washer, SO_2-washer, bottom ash-grating, etc.).

Alternatives

In addition to waste reduction and recycling strategies, there are various alternatives to landfills, including waste-to-energy incineration, anaerobic digestion, composting, mechanical biological treatment, pyrolysis and plasma arc gasification, which have all begun to establish themselves in the market. Depending on local economics and incentives, these can be made more financially attractive than landfills.

Restrictions

Countries including Germany, Austria, Sweden, Denmark, Belgium, the Netherlands, and Switzerland, have banned the disposal of untreated waste in landfills. In these countries, only the ashes from incineration or the stabilized output of mechanical biological treatment plants may still be deposited.

Sanitary Landfill

The term landfill generally refers to an engineered deposit of wastes either in pits/trenches or on the surface. And, a sanitary landfill is essentially a landfill, where proper mechanisms are available to control the environmental risks associated with the disposal of wastes and to make available the land, subsequent to disposal, for other purposes. However, you mustnotethata landfill need not necessarily be an engineered site, when the waste is largely inert at final disposal, as in rural areas, where wastes contain a large proportion of soil and dirt. This practice is generally designated as non-engineered disposal method. When compared to uncontrolled dumping, engineered land-fills are more likely to have pre-planned installations, environmental monitoring, and organised and trained workforce. Sanitary landfill implementation, therefore, requires careful site selection, preparation and management.

The four minimum requirements you need to consider for a sanitary landfill are:

(i)   full or partial hydrological isolation;

(ii) formal engineering preparation;

(iii) permanent control;

(iv) planned waste emplacement and covering.

Against this background, let us now discuss the principles, processes and operation of sanitary landfills.

Principle

The purpose of land filling is to bury or alter the chemical composition of the wastes so that they do not pose any threat to the environment or public health. Landfills are not homogeneous and are usually made up of cells in which a discrete volume of waste is kept isolated from adjacent waste cells by a suitable barrier. The barriersbetween cells generally consist of a layer of natural soil (i.e., clay), which restricts downward or lateral escape of the waste constituents or leachate.

Land filling relies on containment rather than treatment (for control) of wastes. If properly executed, it is a safer and cheaper method than incineration. An environmentally sound sanitary landfill comprises appropriate liners for protection of the groundwater (from contaminated leachate), run-off controls, leachate collection and treatment, monitoring wells and appropriate final cover design (Phelps, 1995).

Figure gives a schematic layout of sanitary landfill along with its various components:

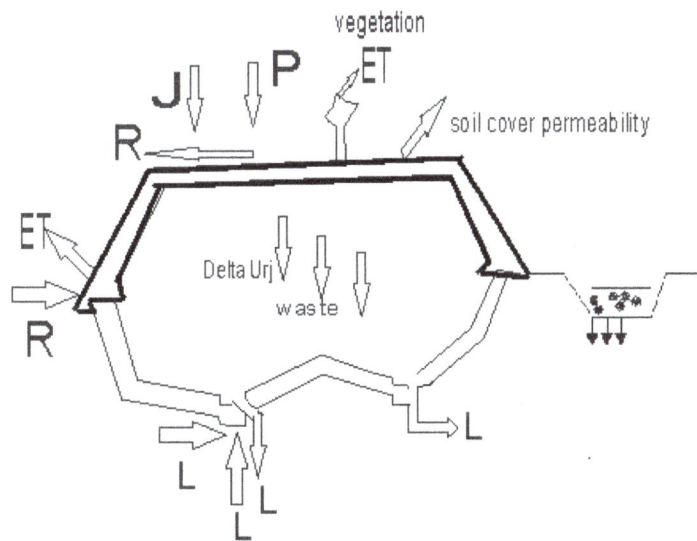

Schematic Layout of Sanitary Landfill

- Planning phase: This typically involves preliminary hydro-geological and geo-technical site investigations as a basis for actual design.

- Construction phase: This involves earthworks, road and facility construction and preparation (liners and drains) of the fill area.

- Operation phase (5 – 20 years): This phase has a high intensity of traffic, work at the front of the fill, operation of environmental installations and completion of finished sections.

- Completed phase (20 – 100 years): This phase involves the termination of the actual filling to the time when the environmental installations need no longer be operated. The emissions may have by then decreased to a level where they do not need any further treatment and can be discharged freely into the surroundings.

- Final storage phase: In this phase, the landfill is integrated into the surroundings for other purposes, and no longer needs special attention.

Landfill Processes

The feasibility of land disposal of solid wastes depends on factors such as the type, quantity and characteristics of wastes, the prevailing laws and regulations, and soil and site characteristics. Let us now explain some of these processes.

(i) Site selection process and considerations: This requires the development of a working plan – a plan, or a series of plans, outlining the development and descriptions of site location, operation, engineering and site restoration. Considerations for site include public opinion, traffic patterns and congestion, climate, zoning requirements, availability of cover material and liner as well, high trees or buffer in the site perimeter, historic buildings, and endangered species, wetlands, and site land environmental factors, speed limits, underpass limitations, load limits on roadways, bridge capacities, and proximity of major roadways, haul distance, hydrology and detours.

(ii) Settling process: The waste body of a landfill undergoes different stagesof settling or deformation. Figure below illustrates these stages:

Figure Settling Processes in Landfill

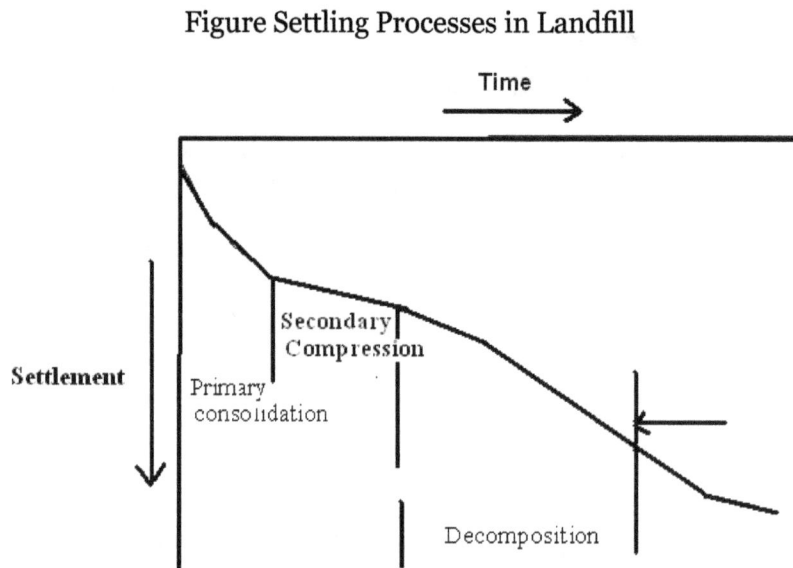

The three stages shown in the figure above are described below:

- Primary consolidation: During this stage, a substantial amount of settling occurs. This settlement is caused by the weight of the waste layers. The movement of trucks, bulldozers or mechanical compactors will also enhance this process. After this primary consolidation, or short-term deformation stage, aerobic degradation processes occur.

- Secondary compression: During this stage, the rate of settling is much lower than that in the primary consolidation stage, as the settling occurs through compression, which cannot be enhanced.

- Decomposition: During the degradation processes, organic material is converted into gas and leachate. The settling rate during this stage increases compared to the secondary compression stage, and continues until all decomposable organic matter is degraded. The settling rate, however, gradually decreases with the passage of time.

To appropriately design protective liners, and gas and leachate collection systems, it is, therefore, necessary to have a proper knowledge of the settling process of wastes.

(iii) Microbial degradation process: The microbial degradation process is the most important biological process occurring in a landfill. These processes induce changes in the chemical and physical environment within the waste body, which determine the quality of leachate and both thequality andquantity of landfill gas. Assuming that landfills mostly receive organic wastes, microbial processes will dominate the stabilisation of the waste and therefore govern landfill gas generation and leachate composition. Soon after disposal, the predominant part of the wastes becomes anaerobic, and the bacteria will start degrading the solid organic carbon, eventually toproduce carbon dioxide and methane. The anaerobic degradation process undergoes the following stages:

- Solid and complex dissolved organic compounds are hydrolysed and fermented by the fermenters primarily to volatile fatty acids, alcohols, hydrogen and carbon dioxide.

- An acidogenic group of bacteria converts the products of the first stage to acetic acid, hydrogen and carbon dioxide.

- Methanogenic bacteria convert acetic acid to methane and carbon dioxide and hydrogenophilic bacteria convert hydrogen and carbon dioxide to methane.

The biotic factors that affect methane formation in the landfill are pH, alkalinity, nutrients, temperature, oxygen and moisture content.

Enhancement of Degradation

Enhancement of the degradation processes in landfills will result in a faster stabilisation of the waste in the landfill, which enhances gas production, and we can achieve this by:

- Adding partly composted waste: As the readily degradable organic matter has already been decomposed aerobically, the rapid acid production phase is overcome, and the balance of acidand methane production bacteria can develop earlier and the consequent dilution effect lowers the organic acid concentration.

- Recirculating leachate: This may have positive effects since a slow increase in moisture will cause a long period of gas production. During warmer periods, recirculated leachate will evaporate, resulting in lower amounts of excess leachate.

Landfill Gas and Leachate

Leachate and landfill gas comprise the major hazards associated with a landfill. While leachate may contaminate the surrounding land and water, landfill gas can be toxic and lead to global warming and explosion leading to human catastrophe (Phelps, 1995). (Note that global warming, also known as greenhouse effect, refers to the warming of the earth's atmosphere by the accumulation of gases (e.g., methane, carbon dioxide and chlorofluorocarbons) that absorbs reflected solar radiation.) The factors, which affect the production of leachate and landfill gas, are the following:

- Nature of waste: The deposition of waste containing biodegradable matter invariably leads to the production of gas and leachate, and the amount depends on the content of biodegradable material in the waste.

- Moisture content: Most micro-organisms require a minimum of approximately 12% (by weight) moisture for growth, and thus the moisture content of landfill waste is an important factor in determining the amount and extent of leachate and gas production.

- pH: The methanogenic bacteria within a landfill produce methane gas, which will grow only at low pH range around neutrality.

- Particle size and density: The size of waste particle affects the density that can be achieved upon compaction and affects the surface area andhence volume. Both affect moisture absorption and therefore are potential for biological degradation.

- Temperature: An increase in temperature tends to increase gas production. The temperature affects the microbial activity to the extent that it is possible to segregate bacteria, according to their optimum temperature operating conditions.

Note that the composition of waste, which varies with region and climate (season), determines the variation in pollution potential. Carbohydrates comprise a large percentage of biodegradable matter within municipal waste, the overall breakdown of which can be represented by the following equation:

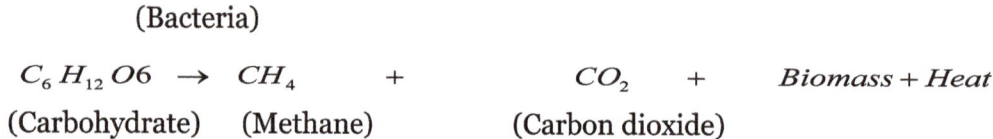

(Bacteria)

$$C_6 H_{12} O6 \rightarrow CH_4 \quad + \quad CO_2 \quad + \quad Biomass + Heat$$

(Carbohydrate)   (Methane)   (Carbon dioxide)

Landfill Gas Emission

Landfill gas contains a high percentage of methane due to the anaerobic decomposition of organic matter, which can be utilised as a source of energy.

Composition and Properties

We can predict the amount and composition of the gas generated for different substrates, depending on the general anaerobic decomposition of wastes added. Climatic and environmental conditions also influence gas composition. Due to the heterogeneous nature of the landfill, some acid-phase anaerobic decomposition occurs along with the methanogenic decomposition.Since aerobic and acid-phase degradation give rise to carbon dioxide and not methane, there may be a higher carbon dioxide content in the gas generated than what would otherwise be expected. Furthermore, depending on the moisture distribution, some carbon dioxide goes into solution. This may appearto increase (artificially) the methane content of the gas measured in the landfill. A typical landfill gas contains a number of components such as the following, which tend to occur within a characteristic range:

- Methane: This is a colourless, odourless and flammable gaswith a density lighter than air, typically making up 50 – 60% of the landfill gas.

- Carbon dioxide: This is a colourless, odourless and non-inflammable gas that is denser than air, typically accounting for 30 – 40%.

- Oxygen: The flammability of methane depends on the percentage of oxygen. It is, therefore, important to control oxygen levels, where gas abstraction is undertaken.

- Nitrogen: This is essentially inert and will have little effect, except to modify the explosive range of methane.

It is difficult to convert the amount of gas measured to the maximum landfill gas production value because gas is withdrawn from a small part of the landfill only, referred to as zone of influence during measurement. In other words, it is very difficult to determine this zone and relate it to the whole landfill area.

Hazards

Landfill gas consists of a mixture of flammable, asphyxiating and noxious gases and may be hazardous to health and safety, and hence the need for precautions. Some of the major hazards are listed below:

- Explosion and fire: Methane is flammable in air within the range of 5 – 15% by volume, while hydrogen is flammable within the range of 4.1 – 7.5% (in the presence of oxygen) and potentially explosive. Fire, occurring within the waste, can be difficult to extinguish and can lead to unpredictable and uncontrolled subsidence as well as production of smoke and toxic fumes.

- Trace components: These comprise mostly alkanes and alkenes, and their oxidation products such as aldehydes, alcohols and esters . Many of them are recognised as toxicants, when present in air at concentrations above occupational exposure standards.

- Global warming: Known also as greenhouse effect, it is the warming of the earth's atmosphere by the accumulation of gases (methane, carbon dioxide and chlorofluorocarbons) that absorbs reflected solar radiation.

Migration

During landfill development, most of the gasproduced isvented to the atmosphere, provided the permeable intermediate cover has been used. While biological and chemical processes affect gas composition through methane oxidation, which converts methane to carbon dioxide, physical factors affect gas migration. The physical factors that affect gas migration include:

- Environmental conditions: These affect the rate of degradation and gas pressure build up.

- Geophysical conditions: These affect migration pathways. In the presence of fractured geological strata or a mineshaft, the gas may travel large distances, unless restricted by the water table.

- Climatic conditions: Falling atmospheric pressure, rainfall and water infiltration rate affect landfill gas migration.

The proportion of void space in the ground, rather than permeability, determines the variability of gas emission. If the escape of landfill gas is controlled and proper extraction system is designed, this gas can be utilised as a source of energy. If landfill gas is not utilised, it should be burnt by means off laring. However, landfill gas utilisation can save on the use of fossil fuels since its heating value is approximately 6 kWh/m3 and can be utilised in internal combustion engines for production of electricity and heat.

It is important that landfill gas is extracted during the operation phase. It is extracted out of the landfill by means of gas wells, which are normally drilled by auger and are driven into the landfill at a spacing of 40 – 70 m. In addition, horizontal systems can be installed during operation of the landfill. The gas wells consist mainly of perforated plastic pipes surrounded by coarse gravel and are connected with the gastransportation pipe with flexible tubing. The vacuum necessary for gas extraction and transportation is created by means of a blower. The most important factors influencing planning and construction of landfill gas extraction systems are settling of waste, water tables in landfills and gas quality.

Control

To control gas emission, it is necessary to control the following:

- waste inputs (i.e., restrict the amount of organic waste).

- processes within the waste (i.e., minimise moisture content to limit gas production).

- migration process (i.e., provide physical barriers or vents to remove the gas from the site and reduce gas pressure). Note that since gas migration cannot be easily prevented, removal is often the preferred option. This is done by using vents (extraction wells) within the waste or stone filled vents, which are often placed around the periphery of the landfill site. Some of the gas collection systems include impermeable cap, granular material, collection pipes and treatment systems.

Leachate Formation

Leachate can pollute both groundwater and surface water supplies. The degree of pollution will depend on local geology and hydrogeology, nature of waste and the proximity of susceptible receptors. Once groundwater is contaminated, it is very costly to clean it up. Landfills, therefore, undergo siting, design and construction procedures that control leachate migration.

Composition and Properties

Leachate comprises soluble components of waste and its degradation products enter water, as it percolates through the landfill. The amount of leachate generated depends on:

- water availability;

- landfill surface condition;

- refuse state;

- condition of surrounding strata.

The major factor, i.e., water availability, is affectedbyprecipitation,surface runoff, waste decomposition and liquid waste disposal. The water balance equation for landfill requires negative or zero ("Lo") so that no excess leachate is produced.This is calculated using the following formula:

$Lo = I - E - aW$

i.e. I – E < aW

where, Lo = free leachate retained at site (equivalent to leachate production minus leachate leaving the site); I = total liquid input;

E = evapotranspiration losses; a = absorption capacity of waste;

W = weight of waste disposed.

Common toxic components in leachate are ammonia and heavy metals, which can be hazardous even at low levels, if they accumulate in the food chain. The presence of ammoniacal nitrogen means that leachate often has to be treated off-site before being discharged to a sewer, since there is no natural bio-chemical path for its removal (Ali, et al., 1995). Leachate composition varies with time and location. Table shows a typical leachate properties and composition at various stages of waste decomposition:

Table:

Properties and Composition of Leachate at Various Stages of Decomposition (mg/l except pH)

Components	Fresh wastes	Aged wastes	Wastes with high moisture
pH	6.2	7.5	8.0
COD	23800	1160	1500
BOD	11900	260	500
TOC	8000	465	450
Volatile acid (as C)	5688	5	12
NH3-N	790	370	1000
NO3-N	3	1	1.0
Ortho-P	0.73	1.4	1.0
Cl	1315	2080	1390
Na	9601	300	1900
Mg	252	185	186
K	780	590	570
Ca	1820	250	158
Mn	27	2.1	0.05
Fe	540	23	2.0
Cu	0.12	0.03	-
Zn	21.5	0.4	0.5
Pb	0.40	0.14	-

Source: Alietal., 1995

Leachate Migration

It is generally difficult to predict the movement of escaped leachate accurately. The main controlling factors are the surrounding geology and hydrogeology. Escape to surface water may be relatively easy to control, but if it escapes to groundwater sources, it can be very difficult both to control and clean up. The degree of groundwater contamination is affected by physical, chemical and biological actions. The relative importance of each process may change, however, if the leachate moves from the landfill to the sub-surface region.

Control

The best way to control leachate is through prevention, which should be integral to the site design. In most cases, it is necessary to control liquid access, collection and treatment, all of which can be done using the following landfill liners:

- Natural liners: These refer to compacted clay or shale, bitumen or soil sealants, etc., and are generally less permeable, resistant to chemical attack and have good sorption properties. They generally do not act as true containment barriers, because sometimes leachate migrates through them.

- Synthetic (geo-membrane) liners: These are typically made up of high or medium density polyethylene and are generally less permeable, easy to install, relatively strong and have good deformation characteristics. They sometimes expand or shrink according to temperature and age.

Note that natural and geo-membrane liners are often combined to enhance the overall efficiency of the containment system. Some of the leachate collection systems include impermeable liner, granular material, collection piping, leachate storage tank; leachate is trucked to a wastewater treatment facility.

Treatment

Concentrations of various substances occurring in leachate are too high to be discharged to surface water or into a sewer system. These concentrations, therefore, have to be reduced by removal, treatment or both. The various treatments of leachate include:

- Leachate recirculation: It is one of the simplest forms of treatment. Recirculation of leachate reduces the hazardous nature of leachate and helps wet the waste, increasing its potential for biological degradation.

- Biological treatment: This removes BOD, ammonia and suspended solids. Leachate from land filled waste can be readily degraded by biological means, due to high content of volatile fatty acids (VFAs). The common methods are aerated lagoons (i.e., special devices which enhance the aerobic process es of degradation of organic substances over the entire depth of the tank) and activated sludge process, which differs from aerated lagoons in that discharged sludge is recirculated and is often used for BOD and ammonia removal. While under conditions of low COD, rotating biological contactors (i.e., biomassis brought into contact with circular blades fixed to a common axle which is rotated) are very effective in removing ammonia. In an anaerobic treatment system, complex organic molecules are fermented in filter. The common types are anaerobic filters, anaerobic lagoon and digesters.

- Physicoche mical treatment: After biological degradation, effluents still contain significant concentrations of different substances. Physicochemical treatment processes could be installed to improve the leachate effluent quality. Some of these processes are flocculation-precipitation.

(Note that addition of chemicals to the water attracts the metal by floc formation). Separation of the floc from water takes place by sedimentation, adsorption and reverse osmosis.

Landfill Operation Issues

Once a potential site has been identified/selected, an assessment of design aspects, including costs for civil works, begins. Important issues to be looked into in this regard are land requirements, types of wastes that are handled, evaluation of seepage potential, design of drainage and seepage control facilities, development of a general operation plan, design of solid waste filling plan and determination of equipment requirements.

With this in view, we will discuss some important factors required for successful implementation and operation of a sanitary landfill in Subsection.

Design and Construction

The design and construction process involves site infrastructure, i.e., the position of the buildings, roads and facilities that are necessary to the efficient running of the site and site engineering, i.e., the basic engineering works needed to shape the site for the reception of wastes and to meet the technical requirements of the working plan (Phelps, 1995). At the outset, however, the potential operator and the licensing authority should agree upon a working plan for the landfill. The disposal license includes the design, earthworks and procedures in the working plan.

What are the processes involved in design and construction? We will study these below:

(i) Site infrastructure: The size, type and number of buildings required at a landfill depend on factors such as the level of waste input, the expected life of the site and environmental factors. Depending on the size and complexity of the landfill, buildings range from single portable cabins to big complexes. However, certain aspects such as the following are common:

o   need to comply with planning, building, fire, healthandsafety regulations and controls;

o   security and resistance to vandalism;

o   durability of service and the possible need to relocate accommodation during the lifetime of the site operations;

o   ease of cleaning and maintenance;

o   availability of services such as electricity, water, drainage and telecommunication.

Paying some attention to the appearance of the site entrance is necessary, as it influences the perception of the public about the landfill site. All landfill sites need to control and keep records of vehicles entering and leaving the site, and have a weighbridge to record waste input data, which can be analysed by a site control office. Note that at small sites, the site control office can be accommodated at the site itself.

(ii) Earthworks: Various features of landfill operations may require substantial earthworks, and therefore, the working plan must include earthworks to be carried out before wastes can be deposited. Details about earthworks gain significance, if artificial liners are to be installed, which involves grading the base and sides of the site (including construction of 25 slopes to drain leachate to the collection areas) and the formation of embankments.Material may also have to be placed in stockpiles for later use at the site. The cell method of operation requires the construction of cell walls. At some sites, it may be necessary to construct earth banks around the site perimeter to screen the landfill operations from the public. Trees or shrubs may then be planted on the banks to enhance the screening effect. The construction of roads leading to disposal sites also involves earthworks.

(iii) Lining landfill sites: Where the use of a liner is envisaged, the suitability of a site for lining should be evaluated at the site investigation stage. However, they should not be installed, until the site has been properly prepared.The areato be lined should be free of objects likely to cause physical damage to the liner, such as vegetation and hard rocks. If synthetic liner materials are used, a binding layer of suitable fine-grained material should be laid to support the liner. However, if the supporting layer consists of low permeable material (e.g., clay), the synthetic liner must be placed on top of this layer. A layer of similar fine- grained material with the thickness of 25 – 30 cm should also be laid above the liner to protect it from subsequent mechanical and environmental damage. During the early phase of operation, particular care should be taken to ensure that the traffic does not damage the liner. Monitoring the quality of groundwater close to the site is necessary to get the feedback on the performance of a liner.

(iv) Leachate and landfill gas management: The basic elements of the leachate collection system (i.e., drain pipes, drainage layers, collection pipes, sumps, etc.) must be installed immediately above the liner, before any waste is deposited. Particular care must also be taken to prevent the drain and collection pipes from settling. During landfill operations, waste cells are covered with soil to avoid additional contact between waste and the environment. The soil layers have to be sufficiently permeable to allow downward leachate transport. Landfill gas is not extracted before completion, which includes construction of final cover, of the waste body. Extraction wells (diameter 0.3 to 1.0 m) may be constructed during or after operation.

(v) Landfill capping: Capping is required to control and minimise leachate generation (by minimising water ingress into the landfill) and facilitate landfill gas control or collection (by installing a low permeability cap over the whole site). A cap may consist of natural (e.g., clay) or synthetic (e.g., poly-ethylene) material with thickness of at least 1 m. An uneven settlement of the waste may be a major cause of cap failure. Designs for capping should, therefore, include consideration of leachate and landfill gas collection wells or vents. For the cap toremain effective, it must be protected from agricultural machinery, drying and cracking, plant root penetration, burrowing animals and erosion.

Operation

To secure public acceptability, landfill operations require careful planning and determination of the extent of environmental effects. The basic factor influencing the planning of site operations is the nature and quantity of incoming wastes. The various aspects of this include the following:

(i) Methods of filling: The following variations in land filling techniques are available (Burner and Kelly, 1972):

- Trench method: This involves the excavation of a trench into which waste is deposited, and the excavated material is then used as cover.

- Area method: Wastes may be deposited in layers and so form terraces over the available area. However, with this type of operation, excessive leachate generation may occur, which may render the control difficult.

- Cell method: This method involves the deposition of wastes within pre- constructed bounded area. It is now the preferred method in the industrialised world, since it encourages the concept of progressive filling and restoration. Operating a cellular method of filling enables wastes to be deposited in a tidy manner, as the cells serve both to conceal the tipping operation and trap much of the litter that has been generated.

- Canyon/depression: This method refers to the placing of suitable wastes against lined canyon or ravine slide slopes. (Slope stability and leachate gas emission control are critical issues for this type of waste placement.)

Commonly Used Land Filling Methods

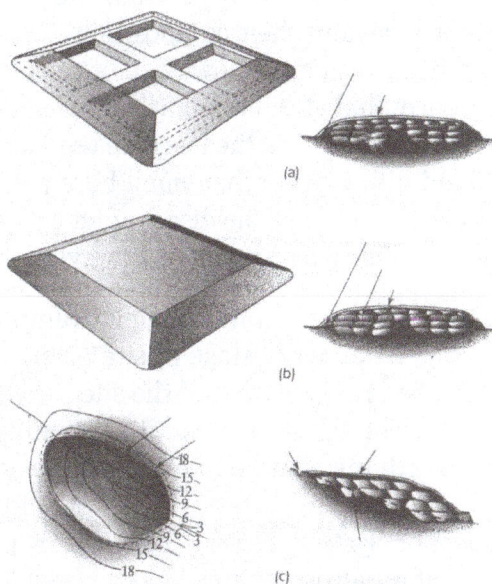

(a) trench (b) areaand (c) canyon/depression methods

(ii) Refuse placement: The working space should be sufficiently extensive to permit vehicles to manoeuvre and unload quickly and safely without impeding refuse spreading, and allow easy operation of the site equipment. Depositing waste in thin layers and using a compactor enables a high waste density to be achieved. Each progressive layer should not be more than 30 cm thick. The number of passes by a machine over the waste determines the level of compaction.

(iii) Covering of waste: At the end of each working day, all exposed surfaces, including the flanks and working space, should be covered witha suitable inert material to a depth of at least 15 cm. This daily cover is considered essential, as it minimises wind blown litter and helps reduce odours. Cover material may be obtained from on-site excavations or inert waste materials coming to the site. Pulverised fuel ash or sewage sludge can also be used for this purpose.

(iv) Site equipment and workforce orientation: The equipment most commonly used on landfill sites includes steel wheeled compactors, tracked dozers, loaders, earthmovers and hydraulic excavators. Scrapers are used for excavating and moving cover materials. In addition to appropriate equipment, proper training must be ensured for the workforce. They should be competent, and adequately supervised; training should include site safetyandfirstaid. Since a landfill site may pose dangers to both site operators and users, it is necessary to lay down emergency plans and test them from time to time (Phelps, 1995).

Monitoring

Landfill represents a complex process of transforming polluting wastes into environmentally acceptable deposits. Because of the complexity of these processes and their potential environmental effects, it is imperative to monitor and confirm that the landfill works, as expected. A monitoring scheme, for example, is required for collecting detailed information on the development of leachate and landfill gas within and beyond a landfill. The scheme should be site specific, drawn at the site investigation stage and implemented. Monitoring is generally done for the following:

(i) Leachate/gas: Monitoring of leachate/gas plays a vital role in the management of landfills. Data on the volume of leachate/gas and their composition are essential for proper control of leachate/gas generation and its treatment. Knowledge of the chemical composition of leachate/gas is also required to confirm that attenuation processes within the landfill are proceeding as expected. Various systems for monitoring the leachate level are in use, and are mostly based on pipes installed prior to land filling. Note that small bore perforated plastic pipes are relatively cheaper and easier to install, but have the disadvantage of getting damaged faster during infilling. Placing pipes within a column or tyres may, however, offer some protection.

(ii) Groundwater: A continued groundwater-monitoring programme for confirming the integrity of the liner system is essential. At an early stage of site preparation, therefore, a number of monitoring boreholes need to be provided around the site. However, the location, design and number of boreholes depend on the size of the landfill, proximity to an aquifer, geology of the site and types of wastes deposited. Installation of a double liner system can make the monitoring exercise more accurate and easier to perform. Water should be regularly flushed through the secondary leachate collection system. In case this water is polluted, the primary leachate barrier will be damaged, and if repair is not considered possible, the leachate collected must be transported to the leachate treatment facility.

Key Issues in Waste Disposal

Let us first get one thing very clear: there is no option but to dispose of wastes. Disposal is the final element in the SWM system. It is the ultimate fate of all solid wastes, be they residential wastes collected and transported directly to a landfill site, semisolid waste (sludge) from municipal and industrial treatment plants, incinerator residue, compost or other substances from various solid waste processing plants that are of no further use to society. It is, therefore, imperative to have a proper plan in place for safe disposal of solid wastes, which involves appropriate handling of residual matter after solid wastes have been processed and the recovery of conversion products/energy has been achieved. It follows that an efficient SWM system must provide an environmentally sound disposal option for waste that cannot be reduced, recycled, composted, combusted, or processed further (Ali, et al 1999).

However, in these days, indiscriminate disposal of wastesin many regionsis very common, giving rise to such problems as:

- health hazards (e.g., residents in the vicinity of wastes inhale dust and smoke when the wastes are burnt; workers and rag pickers come into direct contact with wastes, etc.);

- pollution due to smoke;

- pollution from waste leachate and gas;

- blockage of open drains and sewers.

clearly, safe disposal of solid wastes is important for safeguarding both public health and the environment.

Issues to be Overcome

To achieve effective waste disposal, we must overcome the following the constraints:

(i) Municipal capacities: With the increasing volume of waste generation, collection of wastes gets more attention than disposal. Furthermore, in India, only a few municipalities seem to have the required experience or capacity for controlled disposal. Some municipalities may have identified disposal sites but still only few may actively manage them. In some places, contracting out waste disposal is seen as a solution. But, municipalities are not equipped to deal with the problems associated with it, such as issues of privatisation and monitoring of the contract.

(ii) Political commitment: SWM is more than a technical issue, as any successful programme needs effective political and governmental support. This is rarely a priority of government authorities, unless there is a strong and active public interest as well as international interventions.

(iii) Finance and cost recovery: Development of a sanitary landfill site represents a major investment and it generally receives less priority over other resource demands. And, even when establishment costs are secured for a disposal site, recurrent costs to maintain it always pose problems.

(iv) Technical guidelines: Standards established for waste disposal in one country need not necessarily be appropriate for another, due to reasons such as climatic conditions, resources availability, institutional infrastructure, socio-cultural values, etc. In the absence of adequate data and/or the means of collecting/acquiring it, officials often struggle to plan a safe and economically viable disposal option.

(v) Institutional role and responsibility: A disposal site may be located outside the boundary of a town and may serve more than one town. This necessitates the co-ordination of all authorities concerned, and the roles and responsibilities of different departments need to be clearly defined and accepted by all concerned.

(vi) Location: The accessibility of a disposal site, especially its distance from town, is an important factor in site selection, especially when staff and public do not have a strong incentive to use it, when compared to indiscriminate dumping. Site selection is perhaps the most difficult stage in the development of suitable disposal option.

Solid Waste Treatment: Processes and Technologies

Processing waste requires proper techniques and equipments. It helps in recovering products and energy and in recovering materials for reuse. The process of recycling can help in decreasing consumption which would in return decrease the production of waste. This chapter has been carefully written to provide an easy understanding of the varied facets of solid waste treatment.

Purpose of Processing

The processing of wastes helps in achieving the best possible benefit from every functional element of the solid waste management (SWM) system and, therefore, requires proper selection of techniques and equipment for every element. Accordingly, the wastes that are considered suitable for further use need to be paid special attention in terms of processing, in order that we could derive maximum economical value from them.

The purposes of processing, essentially, are:

(i) Improving efficiency of SWM system: Various processing techniques are available to improve the efficiency of SWM system. For example, before waste papers are reused, they are usually baled to reduce transporting and storage volume requirements. In some cases, wastes are baled to reduce the haul costs at disposal site, where solid wastes are compacted to use the available land effectively. If solid wastesare to be transported hydraulically and pneumatically, some form of shredding is also required. Shredding is also used to improve the efficiency of the disposal site.

(ii) Recovering material for reuse: Usually, materials having a market, when present in wastes in sufficient quantity to justify their separation, are most amenable to recovery and recycling. Materials that can be recovered from solid wastes include paper, cardboard, plastic, glass, ferrous metal, aluminium and other residual metals.

(iii) Recovering conversion products and energy: Combustible organic materials can be converted to intermediate products and ultimatelyto usable energy. This can be done either through incineration, pyrolysis, composting or bio-digestion. Initially, the combustible organic matter is separated from the other solid waste components. Once separated, further processing like shredding and drying is necessary before the waste material can be used for power generation.

Having described the need for waste processing, we now discuss how waste processing is actually carried out.

Mechanical Volume and Size Reduction

Mechanical volume and size reduction is an important factor in the development and operation of any SWM system. The main purpose is to reduce the volume (amount) and size of waste, as compared to its original form, and produce waste of uniform size.

Volume Reduction or Compaction

Volume reduction or compaction refers to densifying wastes in order to reduce their volume. Some of the benefits of compaction include:

- reduction in the quantity of materials to be handled at the disposal site; improved efficiency of collection and disposal of wastes;

- increased life of landfills;

- Economically viable waste management system.

- However, note the following disadvantages associated with compaction:

- poor quality of recyclable materials sorted out of compaction vehicle;

- difficulty in segregation or sorting (since the various recyclable materials are mixed and compressed in lumps);

- Bio-degradable materials (e.g., leftover food, fruits and vegetables) destroy the value of paper and plastic material.

Equipment used for Compaction

Based on their mobility, we can categorise the compaction equipment used in volume reduction under either of the following:

(i) Stationary equipment: This represents the equipment in which wastes are brought to, and loaded into, either manually or mechanically. In fact, the compaction mechanism used to compress waste in a collection vehicle, is a stationary compactor. According to their application, stationary compactors can be described as light duty (e.g., those used for residential areas), commercial or light industrial, heavy industrial and transfer station compactors. Usually, large stationary compactors are necessary, when wastes are to be compressed into:

- steel containers that can be subsequently moved manually or mechanically;

- chambers where the compressed blocks are banded or tied by some means before being removed;

- chambers where they are compressed into a block and then released and hauled away untied;

- transport vehicles directly.

(ii) Movable equipment: This represents the wheeled and tracked equipment used to place and compact solid wastes, as in a sanitary landfill.

Table below lists the types of commonly-used compaction equipment and their suitability:

Types of Compaction Equipment

Location or Operation	Type of Compactor Stationary/residential	Remarks
Solid waste generation points	Vertical	Vertical compaction ram may be used; may be mechanically or hydraulically operated, usually hand-fed; wastes compacted into corrugated box containers, or paper or plastic bags; used in medium and high-rise apartments.
	Rotary	Ram mechanism used to compact waste into paper or plastic bags on rotating platform, platform rotates as containers are filled; used in medium and high-rise apartments.
	Bag or extruder	Compactor can be chute fed; either vertical or horizontal rams; single or continuous multi-bags; single bag must be replaced and continuous bags must be tied off and replaced; used in medium and high-rise apartments.

Location or Operation	Type of Compactor Stationary/residential	Remarks	
	Under counter	Small compactors used in individual residences and apartment units; wastes compacted into special paper bags; after wastes are dropped through a panel door into a bag and door is closed, they are sprayed for odour control; button is pushed to acti vate compaction mechanism.	
	Stationary/commercial	Compactor with vertical and horizontal ram; wastes compressed into steel containers; compressed wastes are manually tied and removed; used in low, medium and high-rise apartments, commercial and industrial facilities.	
Collection	Stationary/packers	Collection vehicles equipped compaction mechanism.	with

Transfer and/ or processing station		Stationary/transfer trailer		Transfer trailer, usually enclosed, equipped with self-contained compaction mechanism.	
	Stationary low pressure		Wastes are compacted into large containers.		
	Stationary high pressure		Wastes are compacted into dense bales or other forms.		
Disposal site		Movable wheeled tracted equipment	or	Specially designed equipment to achieve maximum compaction of wastes.	
	Stationary/track mounted		High-pressure movable stationary compactors used for volume reduction at a disposal site.		

Let us now move on to the discussion of compactors used in the transfer station.

Compactors

According to their compaction pressure, we can divide the compactors used at transfer stations as follows:

(i) Low-pressure (less than 7kg/cm2) compaction: This includes those used at apartments and commercial establishments, bailing equipment used for waste papers and cardboards and stationary compactors used at transfer stations. In low-pressure compaction, wastes are compacted in large containers. Note that portable stationary compactors are being used increasingly by a number of industries in conjunction with material recovery options, especially for waste paper and cardboard.

(ii) High-pressure (more than 7kg/cm2) compaction:Compact systems with

a capacity up to 351.5 kg/cm2 or 5000 lb/in2 come under this category. In such systems, specialised compaction equipment are used to compress solid wastes into blocks or bales of various sizes. In some cases, pulverised wastes are extruded after compaction in the form of logs. The volume reduction achieved with these high-pressure compaction systems varies with the characteristics of the waste. Typically, the reduction ranges from about 3 to 1 through 8 to 1.

When wastes are compressed, their volume is reduced, which is normally expressed in percentage and computed by equation 5.1, given below:

$$\text{VolumeReduction (\%)} = \frac{V_i - V_f}{V_i} \times 100$$

The compaction ratio of the waste is given in equation:

$$\text{Compaction ratio} = \frac{V_i}{V_f}$$

where V_i = volume of waste before compaction, m³ and V_f = volume of waste after compaction, m3

The relationship between the compaction ratio and percent of volume reduction is important in making a trade-off analysis between compaction ratio and cost. Other factors that must be considered are final density of waste after compaction and moisture content. The moisture content that varies with location is another variable that has a major effect on the degree of compaction achieved. In some stationary compactors, provision is made to add moisture, usually in the form of water, during the compaction process.

Selection of Compaction Equipment

To ensure effective processing, we need to consider the following factors, while selecting compaction equipment:

- Characteristics such as size, composition, moisture content, and bulk density of the waste to be compacted.

- Method of transferring and feeding wastes to the compactor, and handling. Potential uses of compacted waste materials.

- Design characteristics such as the size of loading chamber, compaction

- pressure, compaction ratio, etc.

- Operational characteristics such as energy requirements, routine and specialised maintenance requirement, simplicity of operation, reliability, noise output, and air and water pollution control requirement.

- Site consideration, including space and height, access, noise and related environmental limitations.

Size Reduction or Shredding

This is required to convert large sized wastes (as they are collected) into smaller pieces. Size reduction helps in obtaining the final product in a reasonably uniform and considerably reduced size in comparison to the original form. But note that size reduction doesnot necessarily imply volume reduction, and this must be factored in to the design and operation of SWM systems as well as in the recovery of materials for reuse and conversion to energy.

In the overall process of waste treatment and disposal, size reduction is implemented ahead of:

- land filling to provide a more homogeneous product. This may require less cover material and less frequent covering than that without shredding. This can be of economic importance, where cover material is scarce or needs to be brought to the landfill site from some distance.

- recovering materials from the waste stream for recycling.

- baling the wastes – a process sometimes used ahead of long distance transport of solid wastes – to achieve a greater density.

- making the waste a better fuel for incineration waste energy recovery facilities. (The size reduction techniques, coupled with separation techniques such as screening, result in a more homogeneous mixture ofrelatively uniform size, moisture content and heating value, and thereby improving the steps of incineration and energy recovery.

- reducing moisture, i.e., drying and dewatering of wastes.

Equipment used for size reduction

Table: lists the various equipment used for size reduction:

Size Reduction Equipment

Type	Mode of action	Application
Small grinders	Grinding, mashing	Organic residential solid wastes
Chippers	Cutting, slicing	Paper, cardboard, tree trimmings, yard waste, wood, plastics
Large grinders	Grinding, mashing	Brittle and friable materials, used mostly in industrial operation
Jaw crushers	Crushing, breaking	Large solids
Rasp mills	Shredding, tearing	Moistened solid wastes
Shredders	Shearing, tearing	All types of municipal wastes
Cutters, Clippers	Shearing, tearing	All types of municipal wastes
Hammer mills	Breaking, tearing, cutting, crushing	All types of municipal wastes, most commonly used equipment for reducing size and homogenizing composition of wastes
Hydropulper	Shearing, tearing	Ideally suited for use with pulpable wastes, including paper, wood chips. Used primarily in the papermaking industry. Also used to destroy paper records

The most frequently used shredding equipment are the following:

(i) Hammer mill: These are used most often in large commercial operations for reducing the size of wastes. Hammer mill is an impact device consisting of a number of hammers, fastened flexibly to an inner disk, as shown in Figure, which rotates at a very high speed:

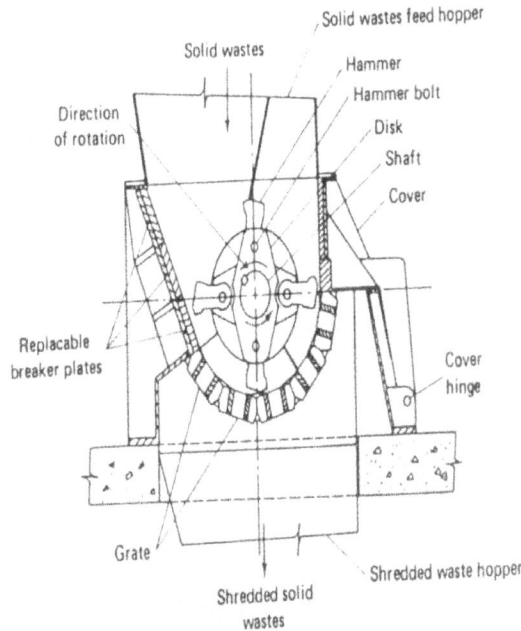

Hammer Mill: An Illustration

Solid wastes, as they enter the mill, are hit by sufficient force, which crush or tear them with a velocity so that they do not adhere to the hammers. Wastes are further reduced in size by being struck between breaker plates and/or cutting bars fixed around the periphery of the inner chamber. This process of cutting and striking action continues, until the required size of material is achieved and after that it falls out of the bottom of the mill.

(ii) Hydropulper: An alternative method of size reduction involves the use of a hydropulper as shown in Figure:

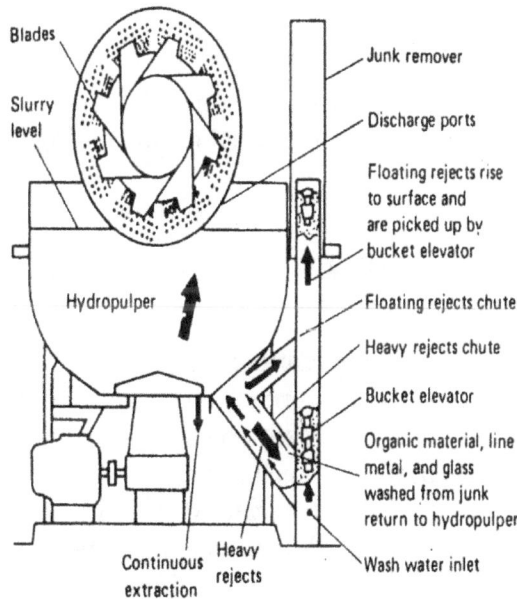

Hydropulper: An Illustration

Solid wastes and recycled water are added to the hydropulper.The high- speed cutting blades, mounted on a rotor in the bottom of the unit, convert pulpable and friable materials into slurry with a solid content varying from2.5 to 3.5%. Metal, tins, cans and other non-pulpable or non-friable materials are rejected from the side of the hydropulper tank. The rejected material passes down a chute that is connected to a bucket elevator, while the solid slurry passes out through the bottom of the pulper tank and is pumped to the next processing operation.

Selection of Size Reduction Equipment

The factorsthat decide the selection of size reduction equipment include the following:

- The properties of materials before and after shredding. Size requirements for shredded material by component.

- Method of feeding shredders, provision of adequate shredder hood capacity

- (toavoidbridging)andclearancerequirementbetweenfeedandtransfer conveyors and shredders.

- Types of operation (continuous or intermittent).

- Operational characteristics including energy requirements, routine and specialised maintenance requirement, simplicity of operation, reliability, noise output, and air and water pollution control requirements.

- Site considerations, including space and height, access, noise and environmental limitations.

- Metal storage after size reduction for the next operation.

Component Separation

Component separation is a necessary operation in which the waste components are identified and sorted either manually or mechanicallytoaidfurther processing. This is required for the:

- recovery of valuable materials for recycling;

- preparation of solid wastes by removing certain components prior to incineration, energy recovery, composting and biogas production.

The most effective way of separation is manual sorting in households prior to collection. In many cities (e.g., Bangalore, Chennai, etc., in India), such systems are now routinely used. The municipality generally provides separate, easily identifiable containers into which the householder deposits segregated recyclable materials such as paper, glass, metals, etc. Usually, separate collections are carried out for the recyclable material. At curbside, separate areas are set aside for each of the recyclable materials for householders to deliver material – when there is no municipal collection system. In case the separation is not done prior to collection, it could be sorted out through mechanical techniques such as air separation, magnetic separation, etc.

Air Separation

This technique has been in use for a number of years in industrial operations for segregating various components from dry mixture. Air separation isprimarily used to separate lighter materials (usually organic) from heavier (usually inorganic) ones. The lighter material mayinclude plastics, paper and paper products and other organic materials. Generally, there is also a need to separate the light fraction of organic material from the conveying air streams, which is usually done in a cyclone separator. In this technique, the heavy fraction is removed from the air classifier (i.e., equipment used for air separation) to the recycling stage or to land disposal, as appropriate. The light fraction may be used, with or without further size reduction, as fuel for incinerators or as compost material. There are various types of air classifiers comm only used,someof which are listed below:

(i) Conventional chutetype: This, as shown in Figure, is one of the simplest types of air classifiers:

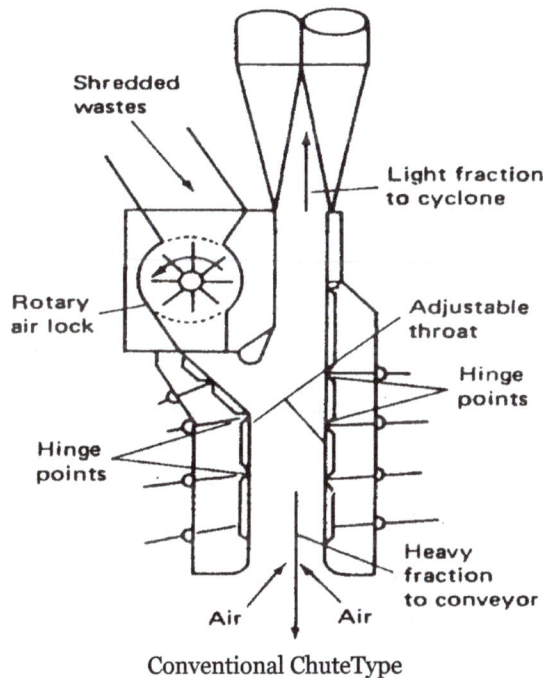

Conventional ChuteType

In this type, when the processed solid wastes are dropped into the vertical chute, the lighter material is carried by the airflow to thetop whilethe heavier materials fall to the bottom of the chute. The control of the percentage split between the light and heavy fraction is accomplished by varying the waste loading rate, airflow rate and the cross section of chute. A rotary air lock feed mechanism is required to introduce theshredded wastes into the classifier.

(ii) Zigzag air classifier: An experimental zigzag air classifier, shown in Figure, consists of a continuous vertical column with internal zigzag deflectors through which air is drawn at a high rate:

Zigzag Air Classifier

Shredded wastes are introduced at the top of the column at a controlled rate, and air is introduced at the bottom of the column. As the wastes drop into the air stream, the lighter fraction is fluidised and moves upward and out of column, while the heavy fraction falls to the bottom. Best separation can be achieved through proper design of the separation chamber, airflow rate and influent feed rate.

(iii) Open inlet vibrator type: Figure below illustrates this type of air classifier:

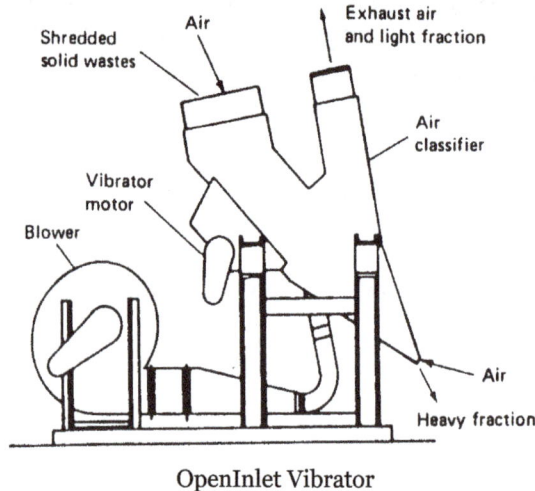

OpenInlet Vibrator

In this type of air classifier, the separation is accomplished bya combination of the following actions:

- Vibration: This helps to stratify the material fed to the separator into heavy and light components. Due to this agitation, the heavier particles tend to settle at the bottom as the shredded waste is conveyed down the length of the separator.

- Inertial force: In this action, the air pulled in through the feed inlet imparts an initial acceleration to the lighter particle, while the wastes travel down the separator as they are being agitated.

- Air pressure: This action refers to the injection of fluidising air in two or more high velocity and low mass flow curtains across the bed. A final stripping of light particles is accomplished at the point where the heavy fraction discharges from the elutriators. It has been reported that the resulting separation is less sensitive to particle size than a conventional vertical air classifier, be it of straight or zigzag design. An advantage ofthis classifier is that an air lock feed mechanismis not required and wastes are fed by gravity directly into the separator inlet.

Selection of Air Separation Equipment

The factors that are to be considered for selecting air separation equipment include the following:

- Characteristics of the material produced by shredding equipment including particle size, shape, moisture content and fibre content.

- Material specification for light fraction.

- Methods of transferring wastes from the shredders to the air separation units and feeding wastes into the air separator.

- Characteristics of separator design including solids-to-air ratio, fluidising velocities, unit capacity, total airflow and pressure drop.

- Operational characteristics including energy requirement, maintenance requirement, simplicity of operation, proved performance and reliability, noise output, and air and water pollution control requirements.

- Site considerations including space and height access, noise and environmental limitations.

So far, we have studied the separation of solid waste components by air separation. We will next learn about the separation of wastes based on their magnetic properties.

Magnetic Separation

The most common method of recovering ferrous scrap from shredded solid wastes involves the use of magnetic recovery systems. Ferrous materials are usually recovered either after shredding or before air classification. When wastes are mass-fired in incinerators, the magnetic separator is used to remove the ferrous material from the incinerator residue. Magnetic recovery systems have also been used at landfill disposal sites. The specific locations, where ferrousmaterials are recovered will depend on the objectives to be achieved, such as reduction of wear and tear on processing and separation equipment, degree of product purity achieved and the required recovery efficiency.

Equipment used for Magnetic Separation

Various types of equipment are in use for the magnetic separation of ferrous materials. The most common types are the following:

(i) Suspended magnet: In this type of separator, a permanent magnet is used to attract the ferrousmetal from the waste stream. When the attracted metal reaches the area, where there is no magnetism, it falls away freely. This ferrous metal is then collected in a container. Figure shows a typical suspended magnet:

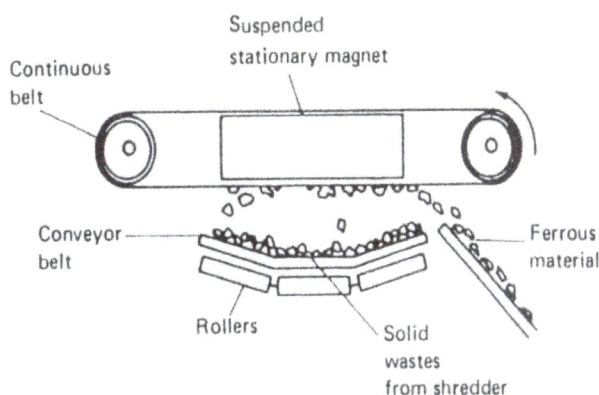

Suspended Type Permanent Magnetic Separator

This type of separation device is suitable for processing raw refuse, where separators can remove large pieces of ferrous metal easily from the waste stream.

(ii) Magnetic pulley: This consists of a drum typedevicecontaining permanent magnets or electromagnets over which a conveyor or a similar transfer mechanism carries the waste stream. The conveyor belt conforms to the rounded shape of the magnetic drum and the magnetic force pulls the ferrous material away from the falling stream of solid waste. Figure illustrates this type of magnetic separator:

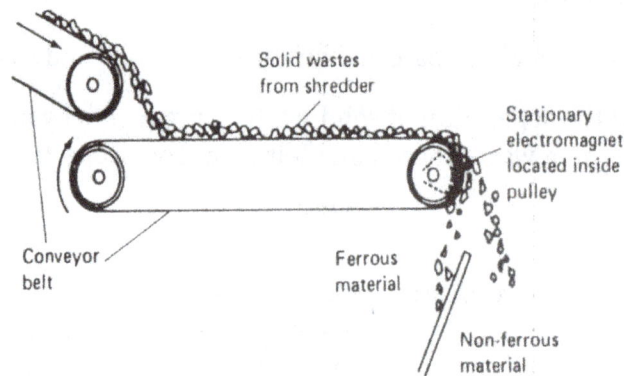

Pulley Type Permanent Magnetic Separator

Selection of Magnetic Separation Equipment

We must consider the following factors in the selection of magnetic separation equipment:

- Characteristics of waste from which ferrous materials are to be separated (i.e., the amount of ferrous material, the tendency of the wastes to stick to each other, size, moisture content, etc.)

- Equipment used for feeding wastes to separator and removing the separated waste streams.

- Characteristics of the separator system engineering design, including loading rate, magnet strength, conveyor speed, material of construction, etc.

- Operational characteristics, including energy requirements, routine and specialised maintenance requirements, simplicity of operation, reliability, noise output, and air and water pollution control requirements.

- Locations where ferrous materials are to be recovered from solid wastes.

- Site consideration, including space and height, access, noise and environmental limitations.

Screening

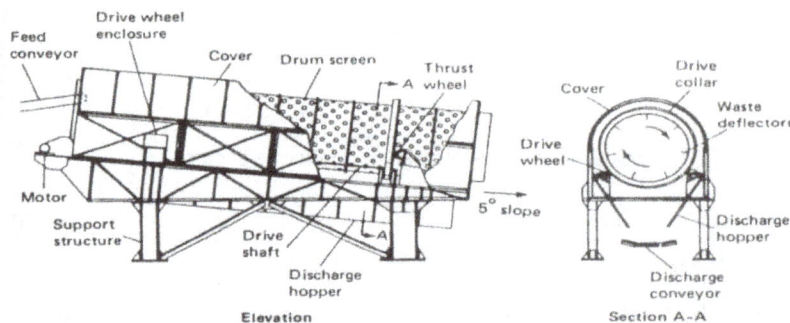

Rotary Drum Screen

Screening is the most common form of separating solid wastes, depending on their size by the use of one or more screening surfaces. Screening has a number of applications in solid waste resource and energy recovery systems. Screens can be used before or after shredding and after air separationofwastesin various applications dealing with both light and heavy fraction materials. The most commonly used screens are rotary drum screens and various forms of vibrating screens.Figures shows a typical rotary drum screen:

Note that rotating wire screens with relatively large openings are used for separation of cardboard and paper products, while vibrating screens and rotating drum screens are typically used for the removal of glass and related materials from the shredded solid wastes.

Selection of Screening Equipment

The various factors that affect the selection of screens include the following:

- Material specification for screened component.

- Location where screening is to be applied and characteristics of waste material to be screened, including particle size, shape, bulk, density and moisture content.

- Separation and overall efficiency.

- Characteristics screen design, including materials of construction, size of screen openings, total surface screening area, oscillating rate for vibrating screens, speed for rotary drum screens, loading rates and length.

- Operational characteristics, including energy requirements, maintenance requirements, simplicity of operation, reliability, noise outputandairand water pollution control requirements.

- Site considerations such as space and height access, noise and related environmental limitations.

The efficiency of screen can be evaluated in terms of the percentage recovery of the material in the feed stream:

$$\text{Recovery (\%)} = \frac{U \times W_u}{F \times W_f} \times 100$$

$$W_f = \frac{Weight\ of\ sample}{Weight\ of\ material\ fed\ to\ the\ screen}$$

$$W_u = \frac{Weight\ of\ sample\ in\ under\ flow}{Total\ weight\ of\ material\ in\ under\ flow}$$

where U = weight of material passing through screen (underflow) kg/h; F

= weight of material fed to the screen, kg/h; W_u = weight fraction of material desired size in underflow; W_f = weight fraction of material of desired size in feed.

The effectiveness of the screening operation can be determined by:

$$Effectiveness = recovery \times rejection$$

where, rejection = 1 − recovery of undesired material

$$= 1 - \frac{U(1 - W_u)}{F(1 - W_f)}$$

Therefore, the effectiveness of screen is:

$$Effectiveness = \frac{U \times W_u}{U \times W_f} \times \left[1 - \frac{U(1 - W_u)}{F(1 - W_f)} \right]$$

Other separation techniques

Besides the mechanical techniques we studied earlier for segregating wastes, there are others. A description of some of these other separation techniques is given below:

(i) Hand-sorting or previewing: Previewing of the waste stream and manual removal of large sized materials is necessary, prior to most types of separation or size reduction techniques. This is done to prevent damage or stoppage of equipment such as shredders or screens, due to items such as rugs, pillows, mattresses, large metallic or plastic objects, wood or other construction materials, paint cans, etc.

(ii) Inertial separation: Inertial methods rely on ballistic or gravity separation principles to separate shredded solid wastes into light (i.e., organic) and heavy (i.e., inorganic) particles. Figures illustrate the modes of operation of two different types of inertial separators:

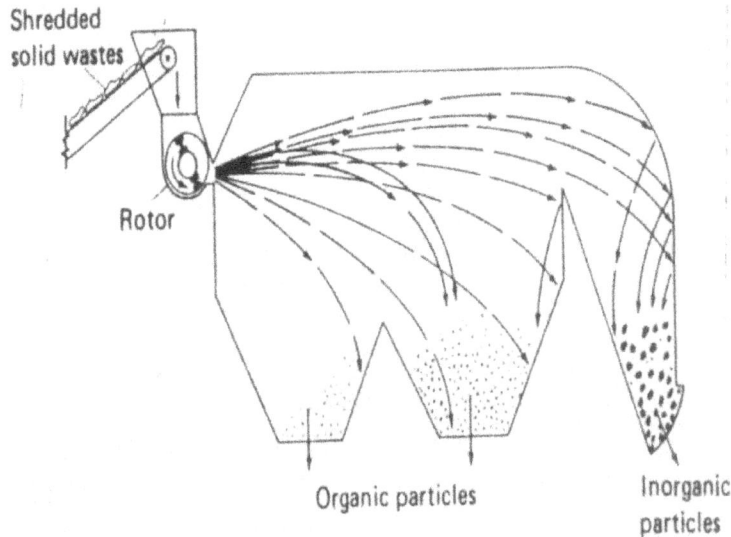

Ballistic Inertial Separator

(iii) Flotation: In the flotation process, glass-rich feedstock, which is produced by screening the heavy fraction of the air-classified wastes after ferrous metal separation, is immersed in water in a solubletank.Glasschips, rocks, bricks, bones and dense plastic materials that sink to the bottom are removed with belt scrappers for further processing. Light organic and othermaterials that float are skimmed from the surface. These materialsare taken to landfill sites or to incinerators for energy recovery. Chemical adhesives (flocculants) are also used to improve the capture of light organic and fine inorganic materials.

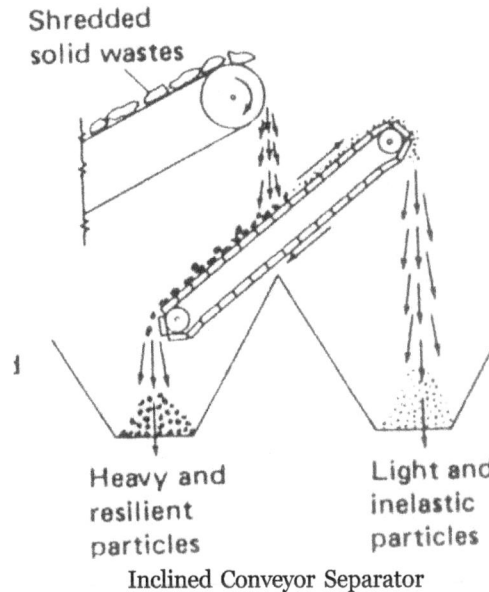

Inclined Conveyor Separator

(iv) Optical sorting: Optical sorting is used mostly to separate glass from the waste stream, and this can be accomplished by identification of the transparent properties of glass to sort it from opaque materials(e.g., stones, ceramics, bottle caps, corks, etc.) in the waste stream. Optical sorting involves a compressed air blast that removes or separates the glasses – plain or coloured. An optical sorting machinery is, however, complex and expensive. Consider Figure shows a simplified scheme of electronic sorter for glass:

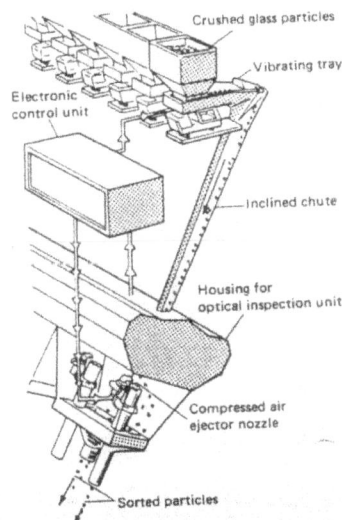

Simplified Scheme of Electronic Sorter

So far, we discussed component separation through air classifiers, magnetic separators, screens, and hand sorting, flotation, optical sorting and inertial separators. In case, however, the waste consists of moisture, we need to remove it for efficient management. It is in this regard that drying and dewatering are considered the most appropriate means of removal of moisture. We will study this next.

Drying and Dewatering

Drying and dewatering operations are used primarily for incineration systems, with or without energy recovery systems. These are also used for drying of sludges in wastewater treatment plants, prior to their incineration or transport to land disposal. The purpose of drying and dewatering operation is to remove moisture from wastes and thereby make it a better fuel. Sometimes, the light fraction is pelletised after drying to make the fuel easier to transport and store, prior to use in an incinerator or energy recovery facility.

Table: shows the range of moisture content for municipal solid waste components:

Moisture Content of Municipal Solid Waste Components

Component	Moisture (in percent)	
	Range	Typical
Food wastes	50 – 80	70
Paper	4 – 10	6
Cardboard	4 – 8	5
Plastics	1 – 4	2
Textiles	6 – 15	10
Rubber	1 – 4	2
Leather	8 – 12	10
Garden trimmings	30 – 80	60
Wood	15 – 40	20
Glass	1 – 4	2
Tin cans	2 – 4	3
Nonferrous metals	2 – 4	2
Ferrous metals	2 – 6	3
Dirt, ashes, brick, etc.	6 – 12	8
Municipal solid wastes	15 – 40	20

Source: Tchobanoglous, et al., (1993)

Drying

The following three methods are used to apply the heat required for drying the wastes:

(i) Convection drying: In this method, hot air is in direct contact with the wet solid waste stream.

(ii) Conduction drying: In this method, the wet solid waste stream is in contact with a heated surface.

(iii) Radiation drying: In this method, heat is transmitted directly to the wet solid waste stream by radiation from the heated body.

Of these three methods, convection drying is used most commonly. Figure illustrates a rotary drum dryer used in the cement industry:

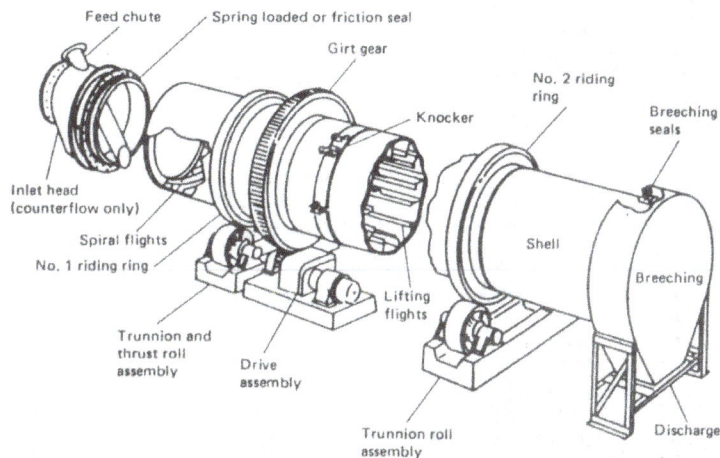

Countercurrent Direct-Heat Rotary Drum Dryer (Bartlett-Snow)

As Figure illustrates, a rotary drum dryer is composed of a rotating cylinder, slightlyinclined from the horizontal through which the material to be dried and the drying gas are passed simultaneously. The drying of material in a direct rotary dryer occurs in the following stages:

- Heating the wet material and its moisture content to the constant-rate drying temperature.

- Drying the material substantially at this temperature.

- Heating of material to its discharge temperature and evaporation of moisture remaining at the end of the stage.

The retention time in the rotary drum is about 30 – 45 minutes. The required energy input will depend on the moisture content, and the required energy input

can be estimated by using a value of about 715 KJ/kg (or 1850 Btu/1b) of water evaporated. Some of the factors, we need to consider in the selection of a drying equipment that include the following:

- Properties of material to be dried.

- Drying characteristics of the materials, including moisture content, maximum material temperature and anticipated drying time.

- Specification of final product, including moisture content. Nature of operation, whether continuous or intermittent.

- Operational characteristics, including energy requirements, maintenance requirements, simplicity of operation, reliability, noise outputandairand water pollution control requirements.

- Site considerations such as space and height access, noise and environmental limitations.

Dewatering

Dewatering is more applicable to the problem of sludge disposal from wastewater treatment of plants, but may also be applicable in some cases to municipal/industrial waste problems. When drying beds, lagoons or spreading on land are not feasible, other mechanical means of dewatering are used. The emphasis in the dewatering operation is often on reducing the liquid volume. Once dewatered, the sludge can be mixed with other solid waste, andthe resulting mixture can be:

- incinerated to reduce volume;

- used for the production of recoverable by-products;

- used for production of compost;

- buried in a landfill.

Centrifugation and filtration are the two common methods for the dewatering of sludge. Sludges with solid content of a few percent can be thickened to about 10– 15% in centrifugation and about 20 – 30% in pressure filtration or vacuum filtration.

Aerobic Treatment System

An aerobic treatment system or ATS, often called (incorrectly) an aerobic septic system, is a small scale sewage treatment system similar to a septic tank system, but which uses an aerobic process for digestion rather than just the anaerobic process used in septic systems. These systems are commonly found in rural areas where public sewers are not available, and may be used for a single residence or for a small group of homes.

Unlike the traditional septic system, the aerobic treatment system produces a high quality secondary effluent, which can be sterilized and used for surface irrigation. This allows much greater flexibility in the placement of the leach field, as well as cutting the required size of the leach field by as much as half.

Process

The ATS process generally consists of the following phases:

- Pre-treatment stage to remove large solids and other undesirable substances.

- Aeration stage, where aerobic bacteria digest biological wastes.

- Settling stage allows undigested solids to settle. This forms a sludge that must be periodically removed from the system.

- Disinfecting stage, where chlorine or similar disinfectant is mixed with the water, to produce an antiseptic output.

The disinfecting stage is optional, and is used where a sterile effluent is required, such as cases where the effluent is distributed above ground. The disinfectant typically used is tablets of calcium hypochlorite, which are specially made for waste treatment systems. The tablets are intended to break down quickly in sunlight. Stabilized forms of chlorine persist after the effluent is dispersed, and can kill plants in the leach field.

Since the ATS contains a living ecosystem of microbes to digest the waste products in the water, excessive amounts of items such as bleach or antibiotics can damage the ATS environment and reduce treatment effectiveness. Non-digestible items should also be avoided, as they will build up in the system and require more frequent sludge removal.

Types of Aerobic Treatment Systems

Small scale aerobic systems generally use one of two designs, fixed-film systems, or continuous flow, suspended growth aerobic systems (CFSGAS). The pre-treatment and effluent handling are similar for both types of systems, and the difference lies in the aeration stage.

Fixed Film Systems

Fixed film systems use a porous medium which provides a bed to support the biomass film that digests the waste material in the wastewater. Designs for fixed film systems vary widely, but fall into two basic categories (though some systems may combine both methods). The first is a system where the media is moved relative to the wastewater, alternately immersing the film and exposing it to air, while the second uses a stationary media, and varies the wastewater flow so the film is alternately submerged and exposed to air. In both cases, the biomass must be exposed to both wastewater and air for the aerobic digestion to occur. The film itself may be made of any suitable porous material, such as formed plastic or peat moss. Simple systems use stationary media, and rely on intermittent, gravity driven wastewater flow to provide periodic exposure to air and wastewater. A common moving media system is the rotating biological contactor (RBC), which uses disks rotating slowly on a horizontal shaft. Approximately 40 percent of the disks are submerged at any given time, and the shaft rotates at a rate of one or two revolutions per minute.

Continuous Flow, Suspended Growth Aerobic Systems

CFSGAS systems, as the name implies, are designed to handle continuous flow, and do not provide a bed for a bacterial film, relying rather on bacteria suspended in the wastewater. The suspension and aeration are typically provided by an air pump, which pumps air through the aeration chamber, providing a constant stirring of the wastewater in addition to the oxygenation. A medium to promote fixed film bacterial growth may be added to some systems designed to handle higher than normal levels of biomass in the wastewater.

Retrofit or Portable Aerobic Systems

Another increasingly common use of aerobic treatment is for the remediation of failing or failed anaerobic septic systems, by retrofitting an existing system with an aerobic feature. This class of product, known as aerobic remediation, is designed to remediate biologically failed and failing anaerobic distribution systems by significantly reducing the biochemical oxygen demand (BOD5) and total suspended solids (TSS) of the effluent. The reduction of the BOD5 and TSS reverses the developed bio-mat. Further, effluent with high dissolved oxygen and aerobic bacteria flow to the distribution component and digest the bio-mat.Doing so on single tank systems where solids do not have anywhere to settle, or there is no a clarifying area can do damage to the field lines as the solid matter is stirred up in the tank.

Composting Toilets

Composting toilets are designed to treat only toilet waste, rather than general residential waste water, and are typically used with water-free toilets rather than the flush toilets associated with the above types of aerobic treatment systems. These systems treat the waste as a moist solid, rather than in liquid suspension, and therefore separate urine from feces during treatment to maintain the correct moisture content in the system. An example of a composting toilet is the clivus multrum (Latin for 'inclined chamber'), which consists of an inclined chamber that separates urine and feces and a fan to provide positive ventilation and prevent odors from escaping through the toilet. Within the chamber, the urine and feces are independently broken down not only by aerobic bacteria, but also by fungi, arthropods, and earthworms. Treatment times are very long, with a minimum time between removals of solid waste of a year; during treatment the volume of the solid waste is decreased by 90 percent, with most being converted into water vapor and carbon dioxide. Pathogens are eliminated from the waste by the long durations in inhospitable conditions in the treatment chamber.

Comparison to Traditional Septic Systems

The aeration stage and the disinfecting stage are the primary differences from a traditional septic system; in fact, an aerobic treatment system can be used as a secondary treatment for septic tank effluent. These stages increase the initial cost of the aerobic system, and also the maintenance requirements over the passive septic system. Unlike many other biofilters, aerobic treatment systems require a constant supply of electricity to drive the air pump increasing overall system costs. The disinfectant tablets must be periodically replaced, as well as the electrical components (air compressor) and mechanical components (air diffusers). On the positive side, an aerobic system produces a higher quality effluent than a septic tank, and thus the leach field can be smaller than that of a conventional septic system, and the output can be discharged in areas too environmentally sensitive for septic system output. Some aerobic systems recycle the effluent through a sprinkler system, using it to water the lawn where regulations approve.

Effluent Quality

Since the effluent from an ATS is often discharged onto the surface of the leach field, the quality is very important. A typical ATS will, when operating correctly, produce an effluent with less than 30 mg/liter BOD5, 25 mg/L TSS, and 10,000 cfu/mL fecal coliform bacteria. This is clean enough that it cannot support a biomat or "slime" layer like a septic tank.

ATS effluent is relatively odorless; a properly operating system will produce effluent that smells musty, but not like sewage. Aerobic treatment is so effective at reducing odors, that it is the preferred method for reducing odor from manure produced by farms.

Trickling Filter

A trickling filter plant in the United Kingdom: The effluent from the primary settling tanks is sprayed onto a bed of coarse gravel (Benfleet Sewage Treatment Plant)

A trickling filter is a type of wastewater treatment system first used by Dibden and Clowes It consists of a fixed bed of rocks, lava, coke, gravel, slag, polyurethanefoam, sphagnum peat moss, ceramic, or plastic media over which sewage or other wastewater flows downward and causes a layer of microbial slime (biofilm) to grow, covering the bed of media. Aerobic conditions are maintained by splashing, diffusion, and either by forced-air flowing through the bed or natural convection of air if the filter medium is porous.

The terms trickle filter, trickling biofilter, biofilter, biological filter and biological trickling filter are often used to refer to a trickling filter. These systems have also been described as roughing filters, intermittent filters, packed media bed filters, alternative septic systems, percolating filters, attached growth processes, and fixed film processes.

Process Description

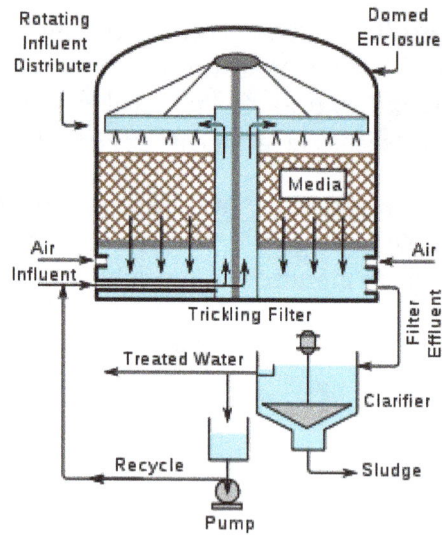

A typical complete trickling filter system

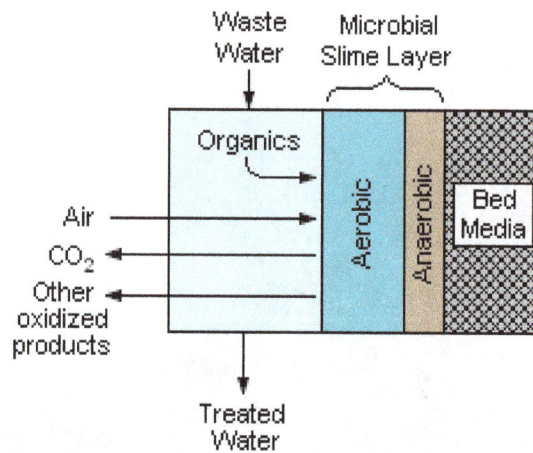

A schematic cross-section of the contact face of the bed of media in a trickling filter

Brocken trickling filter unit at the sewage treatment plant in Norton, Zimbabwe, showing importance of maintenance to prevent structural failure

Typically, sewage flow enters at a high level and flows through the primary settlement tank. The supernatant from the tank flows into a dosing device, often a tipping bucket which delivers flow to the arms of the filter. The flush of water flows through the arms and exits through a series of holes pointing at an angle downwards. This propels the arms around distributing the liquid evenly over the surface of the filter media. Most are uncovered (unlike the accompanying diagram) and are freely ventilated to the atmosphere.

The removal of pollutants from the waste water stream involves both absorption and adsorption of organic compounds and some inorganic species such as nitrite and nitrate ions by the layer of microbial bio film. The filter media is typically chosen to provide a very high surface area to volume. Typical materials are often porous and have considerable internal surface area in addition to the external surface of the medium. Passage of the waste water over the media provides dissolved oxygen which the bio-film layer requires for the biochemicaloxidation of the organic compounds and releases carbon dioxide gas, water and other oxidized end products. As the bio film layer thickens, it eventually sloughs off into the liquid flow and subsequently forms part of the secondary sludge. Typically, a trickling filter is followed by a clarifier or sedimentation tank for the separation and removal of the sloughed film. Other filters utilizing higher-density media such as sand, foam and peat moss do not produce a sludge that must be removed, but require forced air blowers and backwashing or an enclosed anaerobic environment.

Biofilm

The bio-film that develops in a trickling filter may become several millimetres thick and is typically a gelatinous matrix that contains many species of bacteria, cilliates and amoeboid protozoa, annelids, round worms and insect larvae and many other micro fauna. This is very different from many other bio-films which may be less than 1 mm thick. Within the thickness of the biofilm both aerobic and anaerobic zones can exist supporting both oxidative and reductive biological processes. At certain times of year, especially in the spring, rapid growth of organisms in the film may cause the film to be too thick and it may slough off in patches leading to the "spring slough".

Design Considerations

A typical trickling filter is circular and between 10 metres and 20 metres across and between 2 metres to 3 metres deep. A circular wall, often of brick, contains a bed of filter media which in turn rests on a base of under-drains. These under-drains function both to remove liquid passing through the filter media but also to allow the free passage of air up through the filter media. Mounted in the center over the top of the filter media is a spindle supporting two or more horizontal perforated pipes which extend to the edge of the media. The perforations on the pipes are designed to allow an even flow of liquid over the whole area of the media and are also angled so that when liquid flows from the pipes the whole assembly rotates around the central spindle. Settled sewage is delivered to a reservoir at the centre of the spindle via some form of dosing mechanism, often a tipping bucket device on small filters.

Larger filters may be rectangular and the distribution arms may be driven by hydraulic or electrical systems.

Types

Single trickling filters may be used for the treatment of small residential septic tank discharges and very small rural sewage treatment systems. Larger centralized sewage treatment plants typically use many trickling filters in parallel.

Systems can be configured for single-pass use where the treated water is applied to the trickling filter once before being disposed of, or for multi-pass use where a portion of the treated water is cycled back and re-treated via a closed loop. Multi-pass systems result in higher treatment quality and assist in removing Total Nitrogen (TN) levels by promoting nitrification in the aerobic media bed and denitrification in the anaerobic septic tank. Some systems use the filters in two banks operated in series so that the wastewater has two passes through a filter with a sedimentation stage between the two passes. Every few days the filters are switched round to balance the load. This method of treatment can improve nitrification and de-nitrification since much of the carbonaceous oxidative material is removed on the first pass through the filters.

Media Types

Trickling may have a variety of types of filter media used to support the biofilm. Types of media most commonly used include coke, pumice, plastic matrix material, open-cell polyurethanefoam, clinker, gravel, sand and geotextiles. Ideal filter medium optimizes surface area for microbial attachment, wastewater retention time, allows air flow, resists plugging is mechanically robust in all weathers allowing walking access across the filter and does not degrade. Some residential systems require forced aeration units which will increase maintenance and operational costs.

Industrial Wastewater Treatment

The treatment of industrial wastewater may involve specialised tricking filters which use plastic media and high flow rates. Wastewaters from a variety of industrial processes have been treated in trickling filters. Such industrial wastewater trickling filters consist of two types:

- Large tanks or concrete enclosures filled with plastic packing or other media.

- Vertical towers filled with plastic packing or other media.

The availability of inexpensive plastic tower packings has led to their use as trickling filter beds in tall towers, some as high as 20 meters. As early as the 1960s, such towers were in use at: the Great Northern Oil's Pine Bend Refinery in Minnesota; the Cities Service Oil Company Trafalgar Refinery in Oakville, Ontario and at a kraft paper mill.

The treated water effluent from industrial wastewater trickling filters is typically processed in a clarifier to remove the sludge that sloughs off the microbial slime layer attached to the trickling filter media as for other trickling filter applications.

Some of the latest trickle filter technology involves aerated biofilters of plastic media in vessels using blowers to inject air at the bottom of the vessels, with either downflow or upflow of the wastewater.

Sewage Treatment

Wastewater treatment plant in Massachusetts, United States

Sewage treatment is the process of removing contaminants from wastewater, primarily from household sewage. It includes physical, chemical, and biological processes to remove these contaminants and produce environmentally safe treated wastewater (or treated effluent). A by-product of sewage treatment is usually a semi-solid waste or slurry, called sewage sludge, that has to undergo further treatment before being suitable for disposal or land application.

Sewage treatment may also be referred to as wastewater treatment, although the latter is a broader term which can also be applied to purely industrial wastewater. For most cities, the sewer system will also carry a proportion of industrial effluent to the sewage treatment plant which has usually received pretreatment at the factories themselves to reduce the pollutant load. If the sewer system is a combined sewer then it will also carry urban runoff (stormwater) to the sewage treatment plant.

Terminology

The term "sewage treatment plant" (or "sewage treatment works" in some countries) is nowadays often replaced with the term "wastewater treatment plant".

Sewage can be treated close to where the sewage is created, which may be called a "decentralized" system or even an "on-site" system (in septic tanks, biofilters or aerobic treatment systems). Alternatively, sewage can be collected and transported by a network of pipes and pump stations to a municipal treatment plant. This is called a "centralized" system.

Origins of Sewage

Sewage is generated by residential, institutional, commercial and industrial establishments. It includes household waste liquid from toilets, baths, showers, kitchens, and sinks draining into sewers.

In many areas, sewage also includes liquid waste from industry and commerce. The separation and draining of household waste into greywater and blackwater is becoming more common in the developed world, with treated greywater being permitted to be used for watering plants or recycled for flushing toilets.

Sewage Mixing with Rainwater

Sewage may include stormwater runoff or urban runoff. Sewerage systems capable of handling storm water are known as combined sewer systems. This design was common when urban sewerage systems were first developed, in the late 19th and early 20th centuries. Combined sewers require much larger and more expensive treatment facilities than sanitary sewers. Heavy volumes of storm runoff may overwhelm the sewage treatment system, causing a spill or overflow. Sanitary sewers are typically much smaller than combined sewers, and they are not designed to transport stormwater. Backups of raw sewage can occur if excessive infiltration/inflow (dilution by stormwater and/or groundwater) is allowed into a sanitary sewer system. Communities that have urbanized in the mid-20th century or later generally have built separate systems for sewage (sanitary sewers) and stormwater, because precipitation causes widely varying flows, reducing sewage treatment plant efficiency.

As rainfall travels over roofs and the ground, it may pick up various contaminants including soil particles and other sediment, heavy metals, organic compounds, animal waste, and oil and grease. Some jurisdictions require stormwater to receive some level of treatment before being discharged directly into waterways. Examples of treatment processes used for stormwater include retention basins, wetlands, buried vaults with various kinds of media filters, and vortex separators (to remove coarse solids).

Industrial Effluent

In highly regulated developed countries, industrial effluent usually receives at least pretreatment if not full treatment at the factories themselves to reduce the pollutant load, before discharge to the sewer. This process is called industrial wastewater treatment. The same does not apply to many developing countries where industrial effluent is more likely to enter the sewer if it exists, or even the receiving water body, without pretreatment.

Industrial wastewater may contain pollutants which cannot be removed by conventional sewage treatment. Also, variable flow of industrial waste associated with production cycles may upset the population dynamics of biological treatment units, such as the activated sludge process.

Process Steps

Overview

Sewage collection and treatment is typically subject to local, state and federal regulations and standards.

Treating wastewater has the aim to produce an effluent that will do as little harm as possible when discharged to the surrounding environment, thereby preventing pollution compared to releasing untreated wastewater into the environment.

Sewage treatment generally involves three stages, called primary, secondary and tertiary treatment.

- *Primary treatment* consists of temporarily holding the sewage in a quiescent basin where heavy solids can settle to the bottom while oil, grease and lighter solids float to the surface. The settled and floating materials are removed and the remaining liquid may be discharged or subjected to secondary treatment. Some sewage treatment plants that are connected to a combined sewer system have a bypass arrangement after the primary treatment unit. This means that during very heavy rainfall events, the secondary and tertiary treatment systems can be bypassed to protect them from hydraulic overloading, and the mixture of sewage and stormwater only receives primary treatment.

- *Secondary treatment* removes dissolved and suspended biological matter. Secondary treatment is typically performed by indigenous, water-borne micro-organisms in a managed habitat. Secondary treatment may require a separation process to remove the micro-organisms from the treated water prior to discharge or tertiary treatment.

- *Tertiary treatment* is sometimes defined as anything more than primary and secondary treatment in order to allow rejection into a highly sensitive or fragile ecosystem (estuaries, low-flow rivers, coral reefs,...). Treated water is sometimes disinfected chemically or physically (for example, by lagoons and microfiltration) prior to discharge into a stream, river, bay, lagoon or wetland, or it can be used for the irrigation of a golf course, green way or park. If it is sufficiently clean, it can also be used for groundwater recharge or agricultural purposes.

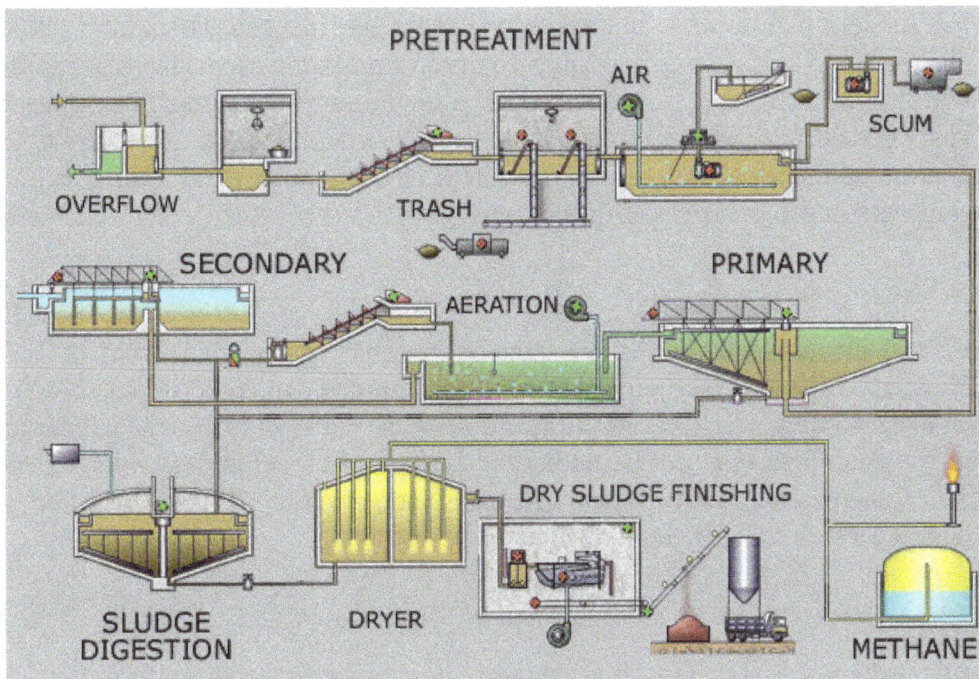

Simplified process flow diagram for a typical large-scale treatment plant

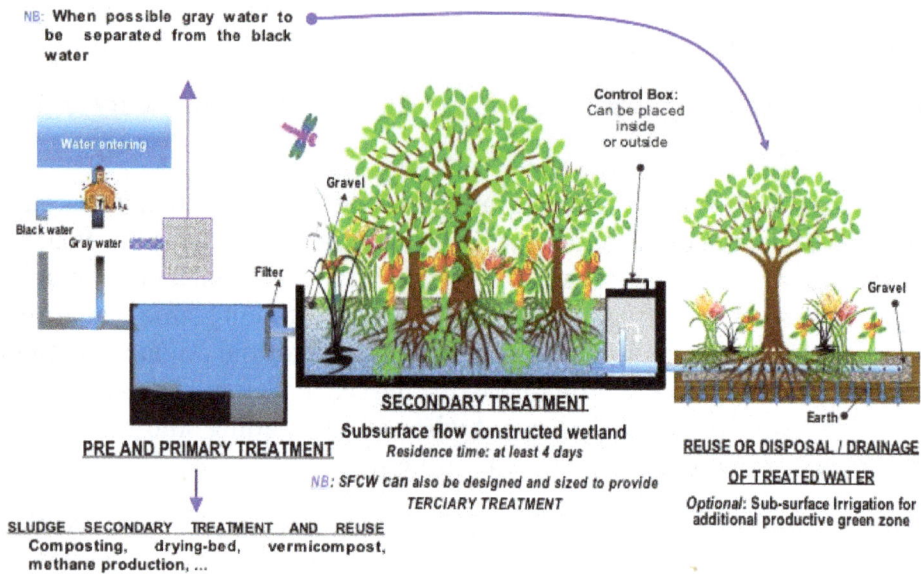

NB: When possible gray water to be separated from the black water

Control Box: Can be placed inside or outside

Water entering

Gravel

Black water
Gray water

Filter

Gravel

SECONDARY TREATMENT
Subsurface flow constructed wetland
Residence time: at least 4 days

Earth

PRE AND PRIMARY TREATMENT

REUSE OR DISPOSAL / DRAINAGE OF TREATED WATER

NB: *SFCW can also be designed and sized to provide*
TERCIARY TREATMENT

Optional: Sub-surface Irrigation for additional productive green zone

SLUDGE SECONDARY TREATMENT AND REUSE
Composting, drying-bed, vermicompost, methane production, ...

Process flow diagram for a typical treatment plant via subsurface flow constructed wetlands (SFCW)

Pretreatment

Pretreatment removes all materials that can be easily collected from the raw sewage before they damage or clog the pumps and sewage lines of primary treatment clarifiers. Objects commonly removed during pretreatment include trash, tree limbs, leaves, branches, and other large objects.

The influent in sewage water passes through a bar screen to remove all large objects like cans, rags, sticks, plastic packets etc. carried in the sewage stream. This is most commonly done with an automated mechanically raked bar screen in modern plants serving large populations, while in smaller or less modern plants, a manually cleaned screen may be used. The raking action of a mechanical bar screen is typically paced according to the accumulation on the bar screens and/or flow rate. The solids are collected and later disposed in a landfill, or incinerated. Bar screens or mesh screens of varying sizes may be used to optimize solids removal. If gross solids are not removed, they become entrained in pipes and moving parts of the treatment plant, and can cause substantial damage and inefficiency in the process.

Grit Removal

Pretreatment may include a sand or grit channel or chamber, where the velocity of the incoming sewage is adjusted to allow the settlement of sand, grit, stones, and broken glass. These particles are removed because they may damage pumps and other equipment. For small sanitary sewer systems, the grit chambers may not be necessary, but grit removal is desirable at larger plants. Grit chambers come in 3 types: horizontal grit chambers, aerated grit chambers and vortex grit chambers. The process is called sedimentation.

Flow Equalization

Clarifiers and mechanized secondary treatment are more efficient under uniform flow conditions. Equalization basins may be used for temporary storage of diurnal or wet-weather flow peaks. Basins provide a place to temporarily hold incoming sewage during plant maintenance and a means of diluting and distributing batch discharges of toxic or high-strength waste which might otherwise inhibit biological secondary treatment (including portable toilet waste, vehicle holding tanks, and septic tank pumpers). Flow equalization basins require variable discharge control, typically include provisions for bypass and cleaning, and may also include aerators. Cleaning may be easier if the basin is downstream of screening and grit removal.

Fat and Grease Removal

In some larger plants, fat and grease are removed by passing the sewage through a small tank where skimmers collect the fat floating on the surface. Air blowers in the base of the tank may also be used to help recover the fat as a froth. Many plants, however, use primary clarifiers with mechanical surface skimmers for fat and grease removal.

Primary Treatment

In the primary sedimentation stage, sewage flows through large tanks, commonly called "pre-settling basins", "primary sedimentation tanks" or "primary clarifiers". The tanks are used to settle sludge while grease and oils rise to the surface and are skimmed off. Primary settling tanks are usually equipped with mechanically driven scrapers that continually drive the collected sludge towards a hopper in the base of the tank where it is pumped to sludge treatment facilities. Grease and oil from the floating material can sometimes be recovered for saponification (soap making).

Primary treatment tanks in Oregon, USA.

Secondary Treatment

Secondary treatment is designed to substantially degrade the biological content of the sewage which are derived from human waste, food waste, soaps and detergent. The majority of municipal plants treat the settled sewage liquor using aerobic biological processes. To be effective, the biota require both oxygen and food to live. The bacteria and protozoa consume biodegradable soluble organic contaminants (e.g. sugars, fats, organic short-chain carbon molecules, etc.) and bind much of the less soluble fractions into floc. Secondary treatment systems are classified as *fixed-film* or *suspended-growth* systems.

- Fixed-film or attached growth systems include trickling filters, bio-towers, and rotating biological contactors, where the biomass grows on media and the sewage passes over its surface. The fixed-film principle has further developed into Moving Bed Biofilm Reactors (MBBR) and Integrated Fixed-Film Activated Sludge (IFAS) processes. An MBBR system typically requires a smaller footprint than suspended-growth systems.

- Suspended-growth systems include activated sludge, where the biomass is mixed with the sewage and can be operated in a smaller space than trickling filters that treat the same amount of water. However, fixed-film systems are more able to cope with drastic changes in the amount of biological material and can provide higher removal rates for organic material and suspended solids than suspended growth systems.

Secondary Sedimentation

Secondary clarifier at a rural treatment plant.

Some secondary treatment methods include a secondary clarifier to settle out and separate biological floc or filter material grown in the secondary treatment bioreactor.

List of Process Types

- Activated sludge

- Aerated lagoon

- Aerobic granulation

- Constructed wetland

- Membrane bioreactor

- Rotating biological contactor

- Sequencing batch reactor

- Trickling filter

To use less space, treat difficult waste, and intermittent flows, a number of designs of hybrid treatment plants have been produced. Such plants often combine at least two stages of the three main treatment stages into one combined stage. In the UK, where a large number of wastewater treatment plants serve small populations, package plants are a viable alternative to building a large structure for each process stage. In the US, package plants are typically used in rural areas, highway rest stops and trailer parks.

Tertiary Treatment

The purpose of tertiary treatment is to provide a final treatment stage to further improve the effluent quality before it is discharged to the receiving environment (sea, river, lake, wet lands, ground, etc.). More than one tertiary treatment process may be used at any treatment plant. If disinfection is practised, it is always the final process. It is also called "effluent polishing."

Filtration

Sand filtration removes much of the residual suspended matter. Filtration over activated carbon, also called *carbon adsorption,* removes residual toxins.

Lagoons or Ponds

A sewage treatment plant and lagoon in Everett, Washington, United States.

Lagoons or ponds provide settlement and further biological improvement through storage in large man-made ponds or lagoons. These lagoons are highly aerobic and colonization by native macrophytes, especially reeds, is often encouraged. Small filter-feeding invertebrates such as *Daphnia* and species of *Rotifera* greatly assist in treatment by removing fine particulates.

Biological Nutrient Removal

Biological nutrient removal (BNR) is regarded by some as a type of secondary treatment process, and by others as a tertiary (or "advanced") treatment process.

Wastewater may contain high levels of the nutrients nitrogen and phosphorus. Excessive release to the environment can lead to a buildup of nutrients, called eutrophication, which can in turn encourage the overgrowth of weeds, algae, and cyanobacteria (blue-green algae). This may cause an algal bloom, a rapid growth in the population of algae. The algae numbers are unsustainable and eventually most of them die. The decomposition of the algae by bacteria uses up so much of the oxygen in the water that most or all of the animals die, which creates more organic matter for the bacteria to decompose. In addition to causing deoxygenation, some algal species produce toxins that contaminate drinking water supplies. Different treatment processes are required to remove nitrogen and phosphorus.

Nitrogen Removal

Nitrogen is removed through the biological oxidation of nitrogen from ammonia to nitrate (nitrification), followed by denitrification, the reduction of nitrate to nitrogen gas. Nitrogen gas is released to the atmosphere and thus removed from the water.

Nitrification itself is a two-step aerobic process, each step facilitated by a different type of bacteria. The oxidation of ammonia (NH_3) to nitrite (NO_2^-) is most often facilitated by *Nitrosomonas* spp. ("nitroso" referring to the formation of a nitroso functional group). Nitrite oxidation to nitrate (NO_3^-), though traditionally believed to be facilitated by *Nitrobacter* spp. (nitro referring the formation of a nitro functional group), is now known to be facilitated in the environment almost exclusively by *Nitrospira* spp.

Denitrification requires anoxic conditions to encourage the appropriate biological communities to form. It is facilitated by a wide diversity of bacteria. Sand filters, lagooning and reed beds can all be used to reduce nitrogen, but the activated sludge process (if designed well) can do the job the most easily. Since denitrification is the reduction of nitrate to dinitrogen (molecular nitrogen) gas, an electron donor is needed. This can be, depending on the waste water, organic matter (from feces), sulfide, or an added donor like methanol. The sludge in the anoxic tanks (denitrification tanks) must be mixed well (mixture of recirculated mixed liquor, return activated sludge [RAS], and raw influent) e.g. by using submersible mixers in order to achieve the desired denitrification.

Sometimes the conversion of toxic ammonia to nitrate alone is referred to as tertiary treatment.

Over time, different treatment configurations have evolved as denitrification has become more sophisticated. An initial scheme, the Ludzack-Ettinger Process, placed an anoxic treatment zone before the aeration tank and clarifier, using the return activated sludge (RAS) from the clarifier as a nitrate source. Influent wastewater (either raw or as effluent from primary clarification) serves as the electron source for the facultative bacteria to metabolize carbon, using the inorganic nitrate as a source of oxygen instead of dissolved molecular oxygen. This denitrification scheme was naturally limited to the amount of soluble nitrate present in the RAS. Nitrate reduction was limited because RAS rate is limited by the performance of the clarifier.

The "Modified Ludzak-Ettinger Process" (MLE) is an improvement on the original concept, for it recycles mixed liquor from the discharge end of the aeration tank to the head of the anoxic tank to provide a consistent source of soluble nitrate for the facultative bacteria. In this instance, raw wastewater continues to provide the electron source, and sub-surface mixing maintains the bacteria in contact with both electron source and soluble nitrate in the absence of dissolved oxygen.

Many sewage treatment plants use centrifugal pumps to transfer the nitrified mixed liquor from the aeration zone to the anoxic zone for denitrification. These pumps are often referred to as *Internal Mixed Liquor Recycle* (IMLR) pumps. IMLR may be 200% to 400% the flow rate of influent wastewater (Q.) This is in addition to Return Activated Sludge (RAS) from secondary clarifiers, which may be 100% of Q. (Therefore, the hydraulic capacity of the tanks in such a system should handle at least 400% of annual average design flow (AADF). At times, the raw or primary effluent wastewater must be carbon-supplemented by the addition of methanol, acetate, or simple food waste (molasses, whey, plant starch) to improve the treatment efficiency. These carbon additions should be accounted for in the design of a treatment facility's organic loading.

Further modifications to the MLE were to come: Bardenpho and Biodenipho processes include additional anoxic and oxidative processes to further polish the conversion of nitrate ion to molecular nitrogen gas. Use of an anaerobic tank following the initial anoxic process allows for luxury uptake of phosphorus by bacteria, thereby biologically reducing orthophosphate ion in the treated wastewater. Even newer improvements, such as Anammox Process, interrupt the formation of nitrate at the nitrite stage of nitrification, shunting nitrite-rich mixed liquor activated sludge to treatment where nitrite is then converted to molecular nitrogen gas, saving energy, alkalinity, and secondary carbon sourcing. Anammox™ (ANaerobic AMMonia OXidation) works by artificially extending detention time and preserving denitrifiying bacteria through the use of substrate added to the mixed liquor and continuously recycled from it prior to secondary clarification. Many other proprietary schemes are being deployed, including DEMON™, Sharon-ANAMMOX™, ANITA-Mox™, and DeAmmon™. The bacteria Brocadia anammoxidans can remove ammonium from waste water through anaerobic oxidation of ammonium to hydrazine, a form of rocket fuel.

Phosphorus Removal

Every adult human excretes between 200 and 1000 grams of phosphorus annually. Studies of United States sewage in the late 1960s estimated mean per capita contributions of 500 grams in urine and feces, 1000 grams in synthetic detergents, and lesser variable amounts used as corrosion and scale control chemicals in water supplies. Source control via alternative detergent formulations has subsequently reduced the largest contribution, but the content of urine and feces will remain unchanged. Phosphorus removal is important as it is a limiting nutrient for algae growth in many fresh water systems. It is also particularly important for water reuse systems where high phosphorus concentrations may lead to fouling of downstream equipment such as reverse osmosis.

Phosphorus can be removed biologically in a process called enhanced biological phosphorus removal. In this process, specific bacteria, called polyphosphate-accumulating organisms (PAOs), are selectively enriched and accumulate large quantities of phosphorus within their cells (up to 20 percent of their mass). When the biomass enriched in these bacteria is separated from the treated water, these biosolids have a high fertilizer value.

Phosphorus removal can also be achieved by chemical precipitation, usually with salts of iron (e.g. ferric chloride), aluminum (e.g. alum), or lime. This may lead to excessive sludge production as hydroxides precipitates and the added chemicals can be expensive. Chemical phosphorus removal requires significantly smaller equipment footprint than biological removal, is easier to operate and is often more reliable than biological phosphorus removal. Another method for phosphorus removal is to use granular laterite.

Once removed, phosphorus, in the form of a phosphate-rich sewage sludge, may be dumped in a landfill or used as fertilizer. In the latter case, the treated sewage sludge is also sometimes referred to as biosolids.

Disinfection

The purpose of disinfection in the treatment of waste water is to substantially reduce the number of microorganisms in the water to be discharged back into the environment for the later use of drinking, bathing, irrigation, etc. The effectiveness of disinfection depends on the quality of the water being treated (e.g., cloudiness, pH, etc.), the type of disinfection being used, the disinfectant dosage (concentration and time), and other environmental variables. Cloudy water will be treated less successfully, since solid matter can shield organisms, especially from ultraviolet light or if contact times are low. Generally, short contact times, low doses and high flows all militate against effective disinfection. Common methods of disinfection include ozone, chlorine, ultraviolet light, or sodium hypochlorite., which is used for drinking water, is not used in the treatment of waste water because of its persistence. After multiple steps of disinfection, the treated water is ready to be released back into the water cycle by means of the nearest body of water or agriculture. Afterwards, the water can be transferred to reserves for everyday human uses.

Chlorination remains the most common form of waste water disinfection in North America due to its low cost and long-term history of effectiveness. One disadvantage is that chlorination of residual organic material can generate chlorinated-organic compounds that may be carcinogenic or harmful to the environment. Residual chlorine or chloramines may also be capable of chlorinating organic material in the natural aquatic environment. Further, because residual chlorine is toxic to aquatic species, the treated effluent must also be chemically dechlorinated, adding to the complexity and cost of treatment.

Ultraviolet (UV) light can be used instead of chlorine, iodine, or other chemicals. Because no chemicals are used, the treated water has no adverse effect on organisms that later consume it, as may be the case with other methods. UV radiation causes damage to the genetic structure of bacteria, viruses, and other pathogens, making them incapable of reproduction. The key disadvantages of UV disinfection are the need for frequent lamp maintenance and replacement and the need for a highly treated effluent to ensure that the target microorganisms are not shielded from the UV radiation (i.e., any solids present in the treated effluent may protect microorganisms from the UV light). In the United Kingdom, UV light is becoming the most common means of disinfection because of the concerns about the impacts of chlorine in chlorinating residual organics in the wastewater and in chlorinating organics in the receiving water. Some sewage treatment systems in Canada and the US also use UV light for their effluent water disinfection.

Ozone (O_3) is generated by passing oxygen (O_2) through a high voltage potential resulting in a third oxygen atom becoming attached and forming O_3. Ozone is very unstable and reactive and oxidizes most organic material it comes in contact with, thereby destroying many pathogenic microorganisms. Ozone is considered to be safer than chlorine because, unlike chlorine which has to be stored on site (highly poisonous in the event of an accidental release), ozone is generated on-site as needed. Ozonation also produces fewer disinfection by-products than chlorination. A disadvantage of ozone disinfection is the high cost of the ozone generation equipment and the requirements for special operators.

Fourth Treatment Stage

Micropollutants such as pharmaceuticals, ingredients of household chemicals, chemicals used in small businesses or industries, environmental persistent pharmaceutical pollutant (EPPP) or pesticides may not be eliminated in the conventional treatment process (primary, secondary and tertiary treatment) and therefore lead to water pollution. Although concentrations of those substances and their decompostion products are quite low, there is still a chance to harm aquatic organisms. For pharmaceuticals, the following substances have been identified as "toxicologically relevant": substances with endocrine disrupting effects, genotoxic substances and substances that enhance the development of bacterial resistances. They mainly belong to the group of environmental persistent pharmaceutical pollutants. Techniques for elimination of micropollutants via a fourth treatment stage during sewage treatment are being tested in Germany, Switzerland and the Netherlands. However, since those techniques are still costly, they are not yet applied on a regular basis. Such process steps mainly consist of activated carbon filters that adsorb the micropollutants. Ozone can also be applied as an oxidative method. Also the use of enzymes such as the enzyme laccase is under investigation. A new concept which could provide an energy-efficient treatment of micropollutants could be the use of laccase secreting fungi cultivated at a wastewater treatment plant to degrade micropollutants and at the same time to provide enzymes at a cathode of a microbial biofuel cells. Microbial biofuel cells are investigated for their property to treat organic matter in wastewater.

To reduce pharmaceuticals in water bodies, also "source control" measures are under investigation, such as innovations in drug development or more responsible handling of drugs.

Odor Control

Odors emitted by sewage treatment are typically an indication of an anaerobic or "septic" condition. Early stages of processing will tend to produce foul-smelling gases, with hydrogen sulfide being most common in generating complaints. Large process plants in urban areas will often treat the odors with carbon reactors, a contact media with bio-slimes, small doses of chlorine, or circulating fluids to biologically capture and metabolize the noxious gases. Other methods of odor control exist, including addition of iron salts, hydrogen peroxide, calcium nitrate, etc. to manage hydrogen sulfide levels.

High-density solids pumps are suitable for reducing odors by conveying sludge through hermetic closed pipework.

Energy Requirements

For conventional sewage treatment plants, around 30 percent of the annual operating costs is usually required for energy. The energy requirements vary with type of treatment process as well as wastewater load. For example, constructed wetlands have a lower energy requirement than activated sludge plants, as less energy is required for the aeration step. Sewage treatment plants that produce biogas in their sewage sludge treatment process with anaerobic digestion can produce enough energy to meet most of the energy needs of the sewage treatment plant itself.

In conventional secondary treatment processes, most of the electricity is used for aeration, pumping systems and equipment for the dewatering and drying of sewage sludge. Advanced wastewater treatment plants, e.g. for nutrient removal, require more energy than plants that only achieve primary or secondary treatment.

Sludge Treatment and Disposal

The sludges accumulated in a wastewater treatment process must be treated and disposed of in a safe and effective manner. The purpose of digestion is to reduce the amount of organic matter and the number of disease-causing microorganisms present in the solids. The most common treatment options include anaerobic digestion, aerobic digestion, and composting. Incineration is also used, albeit to a much lesser degree.

Sludge treatment depends on the amount of solids generated and other site-specific conditions. Composting is most often applied to small-scale plants with aerobic digestion for mid-sized operations, and anaerobic digestion for the larger-scale operations.

The sludge is sometimes passed through a so-called pre-thickener which de-waters the sludge. Types of pre-thickeners include centrifugal sludge thickeners rotary drum sludge thickeners and belt filter presses. Dewatered sludge may be incinerated or transported offsite for disposal in a landfill or use as an agricultural soil amendment.

Environment Aspects

The outlet of the Karlsruhe sewage treatment plant flows into the Alb.

Many processes in a wastewater treatment plant are designed to mimic the natural treatment processes that occur in the environment, whether that environment is a natural water body or the ground. If not overloaded, bacteria in the environment will consume organic contaminants, although this will reduce the levels of oxygen in the water and may significantly change the overall ecology of the receiving water. Native bacterial populations feed on the organic contaminants, and the numbers of disease-causing microorganisms are reduced by natural environmental conditions such as predation or exposure to ultraviolet radiation. Consequently, in cases where the receiving environment provides a high level of dilution, a high degree of wastewater treatment may not be required. However, recent evidence has demonstrated that very low levels of specific contaminants in wastewater, including hormones (from animal husbandry and residue from human hormonal contraception methods) and synthetic materials such as phthalates that mimic hormones in their action, can have an unpredictable adverse impact on the natural biota and potentially on humans if the water is re-used for drinking water. In the US and EU, uncontrolled discharges of wastewater to the environment are not permitted under law, and strict water quality requirements are to be met, as clean drinking water is essential. A significant threat in the coming decades will be the increasing uncontrolled discharges of wastewater within rapidly developing countries.

Effects on Biology

Sewage treatment plants can have multiple effects on nutrient levels in the water that the treated sewage flows into. These nutrients can have large effects on the biological life in the water in contact with the effluent. Stabilization ponds (or sewage treatment ponds) can include any of the following:

- Oxidation ponds, which are aerobic bodies of water usually 1–2 meters in depth that receive effluent from sedimentation tanks or other forms of primary treatment.

    - Dominated by algae

- Polishing ponds are similar to oxidation ponds but receive effluent from an oxidation pond or from a plant with an extended mechanical treatment.

    - Dominated by zooplankton

- Facultative lagoons, raw sewage lagoons, or sewage lagoons are ponds where sewage is added with no primary treatment other than coarse screening. These ponds provide effective treatment when the surface remains aerobic; although anaerobic conditions may develop near the layer of settled sludge on the bottom of the pond.

- Anaerobic lagoons are heavily loaded ponds.

    - Dominated by bacteria

- Sludge lagoons are aerobic ponds, usually 2 to 5 meters in depth, that receive anaerobically digested primary sludge, or activated secondary sludge under water.

    - Upper layers are dominated by algae

Phosphorus limitation is a possible result from sewage treatment and results in flagellate-dominated plankton, particularly in summer and fall.

A phytoplankton study found high nutrient concentrations linked to sewage effluents. High nutrient concentration leads to high chlorophyll a concentrations, which is a proxy for primary production in marine environments. High primary production means high phytoplankton populations and most likely high zooplankton populations, because zooplankton feed on phytoplankton. However, effluent released into marine systems also leads to greater population instability.

The planktonic trends of high populations close to input of treated sewage is contrasted by the bacterial trend. In a study of *Aeromonas* spp. in increasing distance from a wastewater source, greater change in seasonal cycles was found the furthest from the effluent. This trend is so strong that the furthest location studied actually had an inversion of the *Aeromonas* spp. cycle in comparison to that of fecal coliforms. Since there is a main pattern in the cycles that occurred simultaneously at all stations it indicates seasonal factors (temperature, solar radiation, phytoplankton) control of the bacterial population. The effluent dominant species changes from *Aeromonas caviae* in winter to *Aeromonas sobria* in the spring and fall while the inflow dominant species is *Aeromonas caviae*, which is constant throughout the seasons.

Treated Sewage Reuse

With suitable technology, it is possible to reuse sewage effluent for drinking water, although this is usually only done in places with limited water supplies, such as Windhoek and Singapore.

In Israel, about 50 percent of agricultural water use (total use was 1 billion cubic metres in 2008) is provided through reclaimed sewer water. Future plans call for increased use of treated sewer water as well as more desalination plants.

Sewage Treatment in Developing Countries

Few reliable figures exist on the share of the wastewater collected in sewers that is being treated in the world. A global estimate by UNDP and UN-Habitat is that 90% of all wastewater generated is released into the environment untreated. In many developing countries the bulk of domestic and industrial wastewater is discharged without any treatment or after primary treatment only.

In Latin America about 15 percent of collected wastewater passes through treatment plants (with varying levels of actual treatment). In Venezuela, a below average country in South America with respect to wastewater treatment, 97 percent of the country's sewage is discharged raw into the environment. In Iran, a relatively developed Middle Eastern country, the majority of Tehran's population has totally untreated sewage injected to the city's groundwater. However, the construction of major parts of the sewage system, collection and treatment, in Tehran is almost complete, and under development, due to be fully completed by the end of 2012. In Isfahan, Iran's third largest city, sewage treatment was started more than 100 years ago.

Only few cities in sub-Saharan Africa have sewer-based sanitation systems, let alone wastewater treatment plants, an exception being South Africa and – until the late 1990s- Zimbabwe. Instead, most urban residents in sub-Saharan Africa rely on on-site sanitation systems without sewers, such as septic tanks and pit latrines, and faecal sludge management in these cities is an enormous challenge.

History

Basic sewer systems were used for waste removal in ancient Mesopotamia, where vertical shafts carried the waste away into cesspools. Similar systems existed in the Indus Valley civilization in modern-day India and in Ancient Crete and Greece. In the Middle Ages the sewer systems built by the Romans fell into disuse and waste was collected into cesspools that were periodically emptied by workers known as 'rakers' who would often sell it as fertilizer to farmers outside the city.

Modern sewage systems were first built in the mid-nineteenth century as a reaction to the exacerbation of sanitary conditions brought on by heavy industrialization and urbanization. Due to the contaminated water supply, cholera outbreaks occurred in 1832, 1849 and 1855 in London, killing tens of thousands of people. This, combined with the Great Stink of 1858, when the smell of untreated human waste in the River Thames became overpowering, and the report into sanitation reform of the Royal CommissionerEdwin Chadwick, led to the Metropolitan Commission of Sewers appointing Sir Joseph Bazalgette to construct a vast underground sewage system for the safe removal of waste. Contrary to Chadwick's recommendations, Bazalgette's system, and others later built in Continental Europe, did not pump the sewage onto farm land for use as fertilizer; it was simply piped to a natural waterway away from population centres, and pumped back into the environment.

The Great Stink of 1858 stimulated research into the problem of sewage treatment. In this caricature in *The Times*, Michael Faraday reports to *Father Thames* on the state of the river.

Early Attempts

One of the first attempts at diverting sewage for use as a fertilizer in the farm was made by the cotton mill owner James Smith in the 1840s. He experimented with a piped distribution system initially proposed by James Vetch that collected sewage from his factory and pumped it into the outlying farms, and his success was enthusiastically followed by Edwin Chadwick and supported by organic chemist Justus von Liebig.

The idea was officially adopted by the Health of Towns Commission, and various schemes (known as sewage farms) were trialled by different municipalities over the next 50 years. At first, the heavier solids were channeled into ditches on the side of the farm and were covered over when full, but soon flat-bottomed tanks were employed as reservoirs for the sewage; the earliest patent was taken out by William Higgs in 1846 for "tanks or reservoirs in which the contents of sewers and drains from cities, towns and villages are to be collected and the solid animal or vegetable matters therein contained, solidified and dried…" Improvements to the design of the tanks included the introduction of the horizontal-flow tank in the 1850s and the radial-flow tank in 1905. These tanks had to be manually de-sludged periodically, until the introduction of automatic mechanical de-sludgers in the early 1900s.

The precursor to the modern septic tank was the cesspool in which the water was sealed off to prevent contamination and the solid waste was slowly liquified due to anaerobic action; it was invented by L.H Mouras in France in the 1860s. Donald Cameron, as City Surveyor for Exeter patented an improved version in 1895, which he called a 'septic tank'; septic having the meaning of 'bacterial'. These are still in worldwide use, especially in rural areas unconnected to large-scale sewage systems.

Chemical Treatment

Sir Edward Frankland, a distinguished chemist, who demonstrated the possibility of chemically treating sewage in the 1870s.

It was not until the late 19th century that it became possible to treat the sewage by chemically breaking it down through the use of microorganisms and removing the pollutants. Land treatment was also steadily becoming less feasible, as cities grew and the volume of sewage produced could no longer be absorbed by the farmland on the outskirts.

Sir Edward Frankland conducted experiments at the Sewage Farm in Croydon, England, during the 1870s and was able to demonstrate that filtration of sewage through porous gravel produced a nitrified effluent (the ammonia was converted into nitrate) and that the filter remained unclogged over long periods of time. This established the then revolutionary possibility of biological treatment of sewage using a contact bed to oxidize the waste. This concept was taken up by the chief chemist for the London Metropolitan Board of Works, William Libdin, in 1887:

> ...in all probability the true way of purifying sewage...will be first to separate the sludge, and then turn into neutral effluent... retain it for a sufficient period, during which time it should be fully aerated, and finally discharge it into the stream in a purified condition. This is indeed what is aimed at and imperfectly accomplished on a sewage farm.

From 1885 to 1891 filters working on this principle were constructed throughout the UK and the idea was also taken up in the US at the Lawrence Experiment Station in Massachusetts, where Frankland's work was confirmed. In 1890 the LES developed a 'trickling filter' that gave a much more reliable performance.

Contact beds were developed in Salford, Lancashire and by scientists working for the London City Council in the early 1890s. According to Christopher Hamlin, this was part of a conceptual revolution that replaced the philosophy that saw "sewage purification as the prevention of decomposition with one that tried to facilitate the biological process that destroy sewage naturally."

Contact beds were tanks containing the inert substance, such as stones or slate, that maximized the surface area available for the microbial growth to break down the sewage. The sewage was held in the tank until it was fully decomposed and it was then filtered out into the ground. This method quickly became widespread, especially in the UK, where it was used in Leicester, Sheffield, Manchester and Leeds. The bacterial bed was simultaneously developed by Joseph Corbett as Borough Engineer in Salford and experiments in 1905 showed that his method was superior in that greater volumes of sewage could be purified better for longer periods of time than could be achieved by the contact bed.

The Royal Commission on Sewage Disposal published its eighth report in 1912 that set what became the international standard for sewage discharge into rivers; the '20:30 standard', which allowed 20 mg Biochemical oxygen demand and 30 mg suspended solid per litre.

Industrial Wastewater Treatment

Industrial wastewater treatment covers the mechanisms and processes used to treat wastewater that is produced as a by-product of industrial or commercial activities. After treatment, the treated industrial wastewater (or effluent) may be reused or released to a sanitary sewer or to a surface water in the environment. Most industries produce some wastewater although recent trends in the developed world have been to minimise such production or recycle such wastewater within the production process. However, many industries remain dependent on processes that produce wastewaters.

Sources of Industrial Wastewater

Complex Organic Chemicals Industry

A range of industries manufacture or use complex organic chemicals. These include pesticides, pharmaceuticals, paints and dyes, petrochemicals, detergents, plastics, paper pollution, etc. Waste waters can be contaminated by feedstock materials, by-products, product material in soluble or particulate form, washing and cleaning agents, solvents and added value products such as plasticisers. Treatment facilities that do not need control of their effluent typically opt for a type of aerobic treatment, i.e. aerated lagoons.

Electric Power Plants

Fossil-fuel power stations, particularly coal-fired plants, are a major source of industrial wastewater. Many of these plants discharge wastewater with significant levels of metals such as lead, mercury, cadmium and chromium, as well as arsenic, selenium and nitrogen compounds (nitrates and nitrites). Wastewater streams include flue-gas desulfurization, fly ash, bottom ash and flue gas mercury control. Plants with air pollution controls such as wet scrubbers typically transfer the captured pollutants to the wastewater stream.

Ash ponds, a type of surface impoundment, are a widely used treatment technology at coal-fired plants. These ponds use gravity to settle out large particulates (measured as total suspended solids) from power plant wastewater. This technology does not treat dissolved pollutants. Power stations use additional technologies to control pollutants, depending on the particular wastestream in the plant. These include dry ash handling, closed-loop ash recycling, chemical precipitation, biological treatment (such as an activated sludge process), and evaporation.

Food Industry

Wastewater generated from agricultural and food operations has distinctive characteristics that set it apart from common municipal wastewater managed by public or private sewage treatment plants throughout the world: it is biodegradable and non-toxic, but has high concentrations of biochemical oxygen demand (BOD) and suspended solids (SS). The constituents of food and agriculture wastewater are often complex to predict, due to the differences in BOD and pH in effluents from vegetable, fruit, and meat products and due to the seasonal nature of food processing and post-harvesting.

Processing of food from raw materials requires large volumes of high grade water. Vegetable washing generates waters with high loads of particulate matter and some dissolved organic matter. It may also contain surfactants.

Animal slaughter and processing produces very strong organic waste from body fluids, such as blood, and gut contents. This wastewater is frequently contaminated by significant levels of antibiotics and growth hormones from the animals and by a variety of pesticides used to control external parasites.

Processing food for sale produces wastes generated from cooking which are often rich in plant organic material and may also contain salt, flavourings, colouring material and acids or alkali. Very significant quantities of oil or fats may also be present.

Iron and Steel Industry

The production of iron from its ores involves powerful reduction reactions in blast furnaces. Cooling waters are inevitably contaminated with products especially ammonia and cyanide. Production of coke from coal in coking plants also requires water cooling and the use of water in by-products separation. Contamination of waste streams includes gasification products such as benzene, naphthalene, anthracene, cyanide, ammonia, phenols, cresols together with a range of more complex organic compounds known collectively as polycyclic aromatic hydrocarbons (PAH).

The conversion of iron or steel into sheet, wire or rods requires hot and cold mechanical transformation stages frequently employing water as a lubricant and coolant. Contaminants include hydraulic oils, tallow and particulate solids. Final treatment of iron and steel products before onward sale into manufacturing includes *pickling* in strong mineral acid to remove rust and prepare the surface for tin or chromium plating or for other surface treatments such as galvanisation or painting. The two acids commonly used are hydrochloric acid and sulfuric acid. Wastewaters include acidic rinse waters together with waste acid. Although many plants operate acid recovery plants (particularly those using hydrochloric acid), where the mineral acid is boiled away from the iron salts, there remains a large volume of highly acid ferrous sulfate or ferrous chloride to be disposed of. Many steel industry wastewaters are contaminated by hydraulic oil, also known as *soluble oil*.

Mines and Quarries

Mine wastewater effluent in Peru, with neutralized pH from tailing runoff.

The principal waste-waters associated with mines and quarries are slurries of rock particles in water. These arise from rainfall washing exposed surfaces and haul roads and also from rock washing and grading processes. Volumes of water can be very high, especially rainfall related arisings on large sites. Some specialized separation operations, such as coalwashing to separate coal from native rock using density gradients, can produce wastewater contaminated by fine particulate haematite and surfactants. Oils and hydraulic oils are also common contaminants.

Wastewater from metal mines and ore recovery plants are inevitably contaminated by the minerals present in the native rock formations. Following crushing and extraction of the desirable materials, undesirable materials may enter the wastewater stream. For metal mines, this can include unwanted metals such as zinc and other materials such as arsenic. Extraction of high value metals such as gold and silver may generate slimes containing very fine particles in where physical removal of contaminants becomes particularly difficult.

Additionally, the geologic formations that harbour economically valuable metals such as copper and gold very often consist of sulphide-type ores. The processing entails grinding the rock into fine particles and then extracting the desired metal(s), with the leftover rock being known as tailings. These tailings contain a combination of not only undesirable leftover metals, but also sulphide components which eventually form sulphuric acid upon the exposure to air and water that inevitably occurs when the tailings are disposed of in large impoundments. The resulting acid mine drainage, which is often rich in heavy metals (because acids dissolve metals), is one of the many environmental impacts of mining.

Nuclear Industry

The waste production from the nuclear and radio-chemicals industry is dealt with as *Radioactive waste*.

Pulp and Paper Industry

Effluent from the pulp and paper industry is generally high in suspended solids and BOD. Plants that bleach wood pulp for paper making may generate chloroform, dioxins (including 2,3,7,8-TCDD), furans, phenols and chemical oxygen demand (COD). Stand-alone paper mills using imported pulp may only require simple primary treatment, such as sedimentation or dissolved air flotation. Increased BOD or COD loadings, as well as organic pollutants, may require biological treatment such as activated sludge or upflow anaerobic sludge blanket reactors. For mills with high inorganic loadings like salt, tertiary treatments may be required, either general membrane treatments like ultrafiltration or reverse osmosis or treatments to remove specific contaminants, such as nutrients.

Industrial Oil Contamination

Industrial applications where oil enters the wastewater stream may include vehicle wash bays, workshops, fuel storage depots, transport hubs and power generation. Often the wastewater is discharged into local sewer or trade waste systems and must meet local environmental specifications. Typical contaminants can include solvents, detergents, grit. lubricants and hydrocarbons.

Water Treatment

Many industries have a need to treat water to obtain very high quality water for demanding purposes such as environmental discharge compliance. Water treatment produces organic and mineral sludges from filtration and sedimentation. Ion exchange using natural or synthetic resins removes calcium, magnesium and carbonate ions from water, typically replacing them with sodium, chloride, hydroxyl and/or other ions. Regeneration of ion exchange columns with strong acids and alkalis produces a wastewater rich in hardness ions which are readily precipitated out, especially when in admixture with other wastewater constituents.

Wool Processing

Insecticide residues in fleeces are a particular problem in treating waters generated in wool processing. Animal fats may be present in the wastewater, which if not contaminated, can be recovered for the production of tallow or further rendering.

Treatment of Industrial Wastewater

The various types of contamination of wastewater require a variety of strategies to remove the contamination.

Brine Treatment

Brine treatment involves removing dissolved salt ions from the waste stream. Although similarities to seawater or brackish water desalination exist, industrial brine treatment may contain unique combinations of dissolved ions, such as hardness ions or other metals, necessitating specific processes and equipment.

Brine treatment systems are typically optimized to either reduce the volume of the final discharge for more economic disposal (as disposal costs are often based on volume) or maximize the recovery of fresh water or salts. Brine treatment systems may also be optimized to reduce electricity consumption, chemical usage, or physical footprint.

Brine treatment is commonly encountered when treating cooling tower blowdown, produced water from steam assisted gravity drainage (SAGD), produced water from natural gas extraction such as coal seam gas, frac flowback water, acid mine or acid rock drainage, reverse osmosis reject, chlor-alkali wastewater, pulp and paper mill effluent, and waste streams from food and beverage processing.

Brine treatment technologies may include: membrane filtration processes, such as reverse osmosis; ion exchange processes such as electrodialysis or weak acid cation exchange; or evaporation processes, such as brine concentrators and crystallizers employing mechanical vapour recompression and steam.

Reverse osmosis may not be viable for brine treatment, due to the potential for fouling caused by hardness salts or organic contaminants, or damage to the reverse osmosis membranes from hydrocarbons.

Evaporation processes are the most widespread for brine treatment as they enable the highest degree of concentration, as high as solid salt. They also produce the highest purity effluent, even distillate-quality. Evaporation processes are also more tolerant of organics, hydrocarbons, or hardness salts. However, energy consumption is high and corrosion may be an issue as the prime mover is concentrated salt water. As a result, evaporation systems typically employ titanium or duplex stainless steel materials.

Brine Management

Brine management examines the broader context of brine treatment and may include consideration of government policy and regulations, corporate sustainability, environmental impact, recycling, handling and transport, containment, centralized compared to on-site treatment, avoidance and reduction, technologies, and economics. Brine management shares some issues with leachate management and more general waste management.

Solids Removal

Most solids can be removed using simple sedimentation techniques with the solids recovered as slurry or sludge. Very fine solids and solids with densities close to the density of water pose special problems. In such case filtration or ultrafiltration may be required. Although, flocculation may be used, using alum salts or the addition of polyelectrolytes.

Oils and Grease Removal

The effective removal of oils and grease is dependent on the characteristics of the oil in terms of its suspension state and droplet size, which will in turn affect the choice of separator technology.

Oil pollution in water usually comes in four states, often in combination:

- free oil - large oil droplets sitting on the surface;

- heavy oil, which sits at the bottom, often adhering to solids like dirt;

- emulsified, where the oil droplets are heavily "chopped"; and

- dissolved oil, where the droplets are fully dispersed and not visible. Emulsified oil droplets are the most common in industrial oily wastewater and are extremely difficult to separate.

The methodology for separating the oil is dependent on the oil droplet size. Larger oil droplets such as those in free oil pollution are easily removed, but as the droplets become smaller, some separator technologies perform better than others.

Most separator technologies will have an optimum range of oil droplet sizes that can be effectively treated. This is known as the "micron rating."

Analysing the oily water to determine droplet size can be performed with a video particle analyser. Alternatively, there are commonalities in industries for oil droplet sizes. Larger droplets—greater than 60 microns—are often present in wastewater in workshops, re-fuel areas and depots. Twenty to 50 micron oil droplets often are present in vehicle wash bays, meat processing and dairy manufacturing effluent and aluminium billet cooling towers. Smaller droplets in the range of 10 to 20 microns tend to occur in workshops and condensates.

Each separator technology will have its' own performance curve outlining optimum performance based on oil droplet size. the most common separators are gravity tanks or pits, API oil-water separators or plate packs, chemical treatment via DAFs, centrifuges, media filters and hydrocyclones.

API Separators

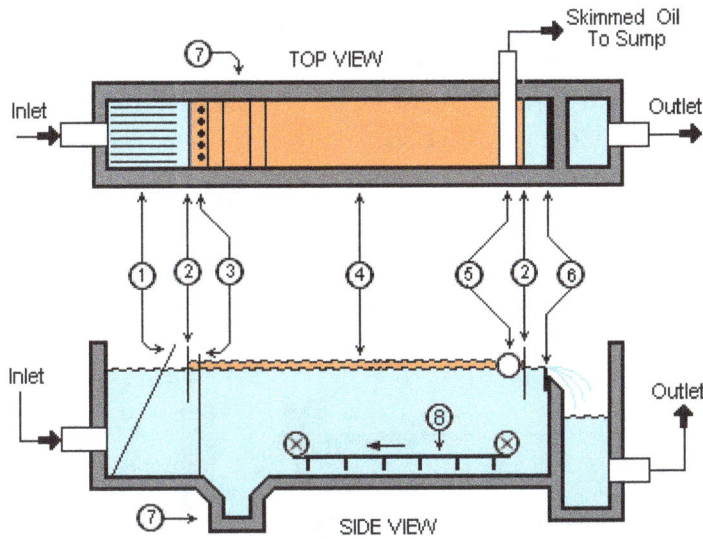

TOP VIEW

Skimmed Oil
To Sump

Inlet

Outlet

SIDE VIEW

Inlet

Outlet

1 Trash trap (inclined rods)
2 Oil retention baffles
3 Flow distributors (vertical rods)
4 Oil layer
5 Slotted pipe skimmer
6 Adjustable overflow weir
7 Sludge sump
8 Chain and flight scraper

A typical API oil-water separator used in many industries

Many oils can be recovered from open water surfaces by skimming devices. Considered a dependable and cheap way to remove oil, grease and other hydrocarbons from water, oil skimmers can sometimes achieve the desired level of water purity. At other times, skimming is also a cost-efficient method to remove most of the oil before using membrane filters and chemical processes. Skimmers will prevent filters from blinding prematurely and keep chemical costs down because there is less oil to process.

Because grease skimming involves higher viscosity hydrocarbons, skimmers must be equipped with heaters powerful enough to keep grease fluid for discharge. If floating grease forms into solid clumps or mats, a spray bar, aerator or mechanical apparatus can be used to facilitate removal.

However, hydraulic oils and the majority of oils that have degraded to any extent will also have a soluble or emulsified component that will require further treatment to eliminate. Dissolving or emulsifying oil using surfactants or solvents usually exacerbates the problem rather than solving it, producing wastewater that is more difficult to treat.

The wastewaters from large-scale industries such as oil refineries, petrochemical plants, chemical plants, and natural gas processing plants commonly contain gross amounts of oil and suspended solids. Those industries use a device known as an API oil-water separator which is designed to separate the oil and suspended solids from their wastewater effluents. The name is derived from the fact that such separators are designed according to standards published by the American Petroleum Institute (API).

The API separator is a gravity separation device designed by using Stokes Law to define the rise velocity of oil droplets based on their density and size. The design is based on the specific gravity difference between the oil and the wastewater because that difference is much smaller than the specific gravity difference between the suspended solids and water. The suspended solids settles to the bottom of the separator as a sediment layer, the oil rises to top of the separator and the cleansed wastewater is the middle layer between the oil layer and the solids.

Typically, the oil layer is skimmed off and subsequently re-processed or disposed of, and the bottom sediment layer is removed by a chain and flight scraper (or similar device) and a sludge pump. The water layer is sent to further treatment for additional removal of any residual oil and then to some type of biological treatment unit for removal of undesirable dissolved chemical compounds.

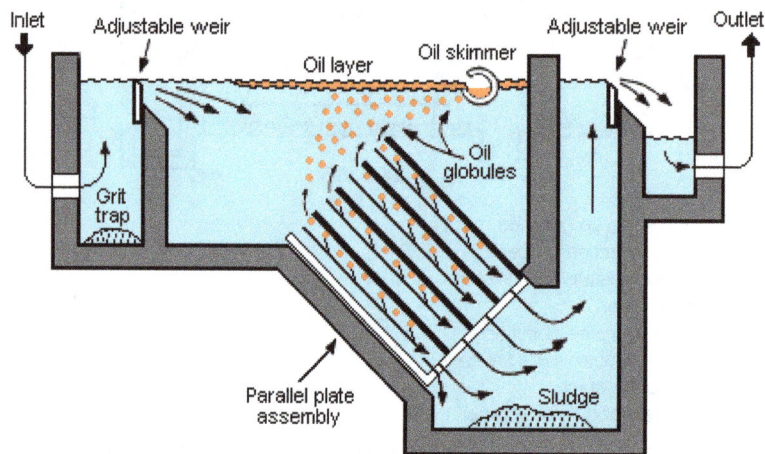

A typical parallel plate separator

Parallel plate separators are similar to API separators but they include tilted parallel plate assemblies (also known as parallel packs). The parallel plates provide more surface for suspended oil droplets to coalesce into larger globules. Such separators still depend upon the specific gravity between the suspended oil and the water. However, the parallel plates enhance the degree of oil-water separation. The result is that a parallel plate separator requires significantly less space than a conventional API separator to achieve the same degree of separation.

Hydrocyclone Oil Separators

Hydrocyclone oil separators operate on the process where wastewater enters the cyclone chamber and is spun under extreme centrifugal forces more than 1000 times the force of gravity. This force causes the water and oil droplets to separate. The separated oil is discharged from one end of the cyclone where treated water is discharged through the opposite end for further treatment, filtration or discharge.

Hydrocyclones are useful for the greatest range of oil droplet sizes operating from less than 10 microns and up and can operate continuously without water pre-treatment and at any temperature and pH. Applications where hydrocyclones are found are in industry where oily water sources arise in workshops, vehicle wash bays, transport hubs, fuel depots and aluminium billet processing. Animal fats from meat processing and dairy manufacturing can also be removed without the need of chemical treatment that often is required for dissolved air flotation (DAF) systems.

Removal of Biodegradable Organics

Biodegradable organic material of plant or animal origin is usually possible to treat using extended conventional sewage treatment processes such as activated sludge or trickling filter. Problems can arise if the wastewater is excessively diluted with washing water or is highly concentrated such as undiluted blood or milk. The presence of cleaning agents, disinfectants, pesticides, or antibiotics can have detrimental impacts on treatment processes.

Activated Sludge Process

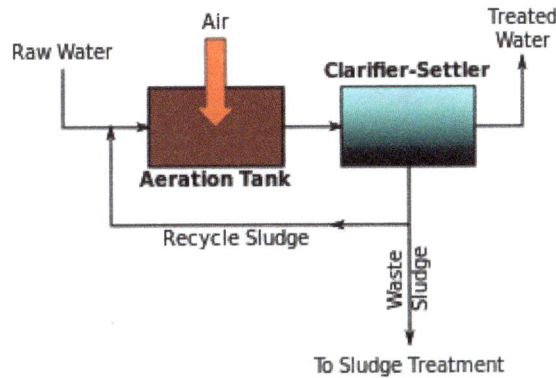

A generalized diagram of an activated sludge process.

Activated sludge is a biochemical process for treating sewage and industrial wastewater that uses air (or oxygen) and microorganisms to biologically oxidize organic pollutants, producing a waste sludge (or floc) containing the oxidized material. In general, an activated sludge process includes:

- An aeration tank where air (or oxygen) is injected and thoroughly mixed into the wastewater.

- A settling tank (usually referred to as a clarifier or "settler") to allow the waste sludge to settle. Part of the waste sludge is recycled to the aeration tank and the remaining waste sludge is removed for further treatment and ultimate disposal.

Trickling Filter Process

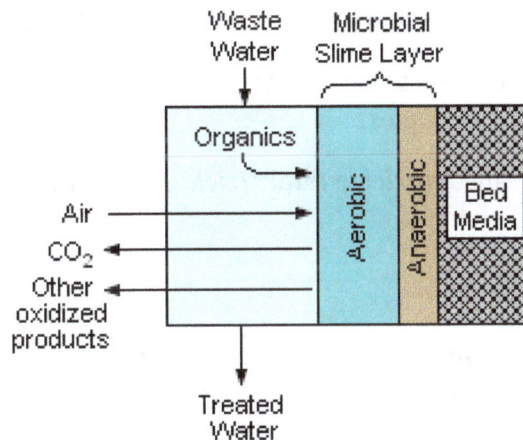

A schematic cross-section of the contact face of the bed media in a trickling filter

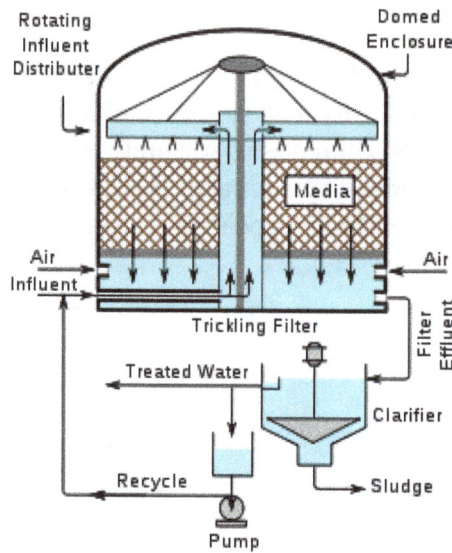

A typical complete trickling filter system

A trickling filter consists of a bed of rocks, gravel, slag, peat moss, or plastic media over which wastewater flows downward and contacts a layer (or film) of microbial slime covering the bed media. Aerobic conditions are maintained by forced air flowing through the bed or by natural convection of air. The process involves adsorption of organic compounds in the wastewater by the microbial slime layer, diffusion of air into the slime layer to provide the oxygen required for the biochemical oxidation of the organic compounds. The end products include carbon dioxide gas, water and other products of the oxidation. As the slime layer thickens, it becomes difficult for the air to penetrate the layer and an inner anaerobic layer is formed.

The fundamental components of a complete trickling filter system are:

- A bed of filter medium upon which a layer of microbial slime is promoted and developed.

- An enclosure or a container which houses the bed of filter medium.

- A system for distributing the flow of wastewater over the filter medium.

- A system for removing and disposing of any sludge from the treated effluent.

The treatment of sewage or other wastewater with trickling filters is among the oldest and most well characterized treatment technologies.

A trickling filter is also often called a *trickle filter, trickling biofilter, biofilter, biological filter* or *biological trickling filter*.

Treatment of Other Organics

Synthetic organic materials including solvents, paints, pharmaceuticals, pesticides, products from coke production and so forth can be very difficult to treat. Treatment methods are often specific to the material being treated. Methods include advanced oxidation processing, distillation, adsorption, vitrification, incineration, chemical immobilisation or landfill disposal. Some materials such as some detergents may be capable of biological degradation and in such cases, a modified form of wastewater treatment can be used.

Treatment of Acids and Alkalis

Acids and alkalis can usually be neutralised under controlled conditions. Neutralisation frequently produces a precipitate that will require treatment as a solid residue that may also be toxic. In some cases, gases may be evolved requiring treatment for the gas stream. Some other forms of treatment are usually required following neutralisation.

Waste streams rich in hardness ions as from de-ionisation processes can readily lose the hardness ions in a buildup of precipitated calcium and magnesium salts. This precipitation process can cause severe *furring* of pipes and can, in extreme cases, cause the blockage of disposal pipes. A 1 metre diameter industrial marine discharge pipe serving a major chemicals complex was blocked by such salts in the 1970s. Treatment is by concentration of de-ionisation waste waters and disposal to landfill or by careful pH management of the released wastewater.

Treatment of Toxic Materials

Toxic materials including many organic materials, metals (such as zinc, silver, cadmium, thallium, etc.) acids, alkalis, non-metallic elements (such as arsenic or selenium) are generally resistant to biological processes unless very dilute. Metals can often be precipitated out by changing the pH or by treatment with other chemicals. Many, however, are resistant to treatment or mitigation and may require concentration followed by landfilling or recycling. Dissolved organics can be *incinerated* within the wastewater by the advanced oxidation process.

Agricultural Wastewater Treatment

Riparian buffer lining a creek in Iowa

Agricultural wastewater treatment is a farm management agenda for controlling pollution from surface runoff that may be contaminated by chemicals in fertiliser, pesticides, animal slurry, crop residues or irrigation water.

Nonpoint Source Pollution

Nonpoint source pollution from farms is caused by surface runoff from fields during rain storms. Agricultural runoff is a major source of pollution, in some cases the only source, in many watersheds.

Sediment Runoff

Highly erodible soils on a farm in Iowa

Soil washed off fields is the largest source of agricultural pollution in the United States. Excess sediment causes high levels of turbidity in water bodies, which can inhibit growth of aquatic plants, clog fishgills and smother animal larvae.

Farmers may utilize erosion controls to reduce runoff flows and retain soil on their fields. Common techniques include:

- contour ploughing
- crop mulching
- crop rotation
- planting perennial crops
- installing riparian buffers.

Nutrient Runoff

Manure spreader

Nitrogen and phosphorus are key pollutants found in runoff, and they are applied to farmland in several ways, such as in the form of commercial fertilizer, animal manure, or municipal or industrial wastewater (effluent) or sludge. These chemicals may also enter runoff from crop residues, irrigation water, wildlife, and atmospheric deposition.

Farmers can develop and implement nutrient management plans to mitigate impacts on water quality by:

- mapping and documenting fields, crop types, soil types, water bodies

- developing realistic crop yield projections

- conducting soil tests and nutrient analyses of manures and/or sludges applied

- identifying other significant nutrient sources (e.g., irrigation water)

- evaluating significant field features such as highly erodible soils, subsurface drains, and shallow aquifers

- applying fertilizers, manures, and/or sludges based on realistic yield goals and using precision agriculture techniques.

Pesticides

Aerial application (crop dusting) of pesticides over a soybean field in the U.S.

Pesticides are widely used by farmers to control plant pests and enhance production, but chemical pesticides can also cause water quality problems. Pesticides may appear in surface water due to:

- direct application (e.g. aerial spraying or broadcasting over water bodies)

- runoff during rain storms

- aerial drift (from adjacent fields).

Some pesticides have also been detected in groundwater.

Farmers may use Integrated Pest Management (IPM) techniques (which can include biological pest control) to maintain control over pests, reduce reliance on chemical pesticides, and protect water quality.

There are few safe ways of disposing of pesticide surpluses other than through containment in well managed landfills or by incineration. In some parts of the world, spraying on land is a permitted method of disposal.

Point Source Pollution

Farms with large livestock and poultry operations, such as factory farms, can be a major source of point source wastewater. In the United States, these facilities are called *concentrated animal feeding operations* or *confined animal feeding operations* and are being subject to increasing government regulation.

Animal Wastes

Confined Animal Feeding Operation in the United States

The constituents of animal wastewater typically contain

- Strong organic content — much stronger than human sewage

- High solids concentration

- High nitrate and phosphorus content

- Antibiotics

- Synthetic hormones

- Often high concentrations of parasites and their eggs

- Spores of *Cryptosporidium* (a protozoan) resistant to drinking water treatment processes

- Spores of *Giardia*

- Human pathogenic bacteria such as *Brucella* and *Salmonella*

Animal wastes from cattle can be produced as solid or semisolid manure or as a liquid slurry. The production of slurry is especially common in housed dairy cattle.

Treatment

Whilst solid manure heaps outdoors can give rise to polluting wastewaters from runoff, this type of waste is usually relatively easy to treat by containment and/or covering of the heap.

Animal slurries require special handling and are usually treated by containment in lagoons before disposal by spray or trickle application to grassland. Constructed wetlands are sometimes used to facilitate treatment of animal wastes, as are anaerobic lagoons. Excessive application or application to sodden land or insufficient land area can result in direct runoff to watercourses, with the potential for causing severe pollution. Application of slurries to land overlying aquifers can result in direct contamination or, more commonly, elevation of nitrogen levels as nitrite or nitrate.

The disposal of any wastewater containing animal waste upstream of a drinking water intake can pose serious health problems to those drinking the water because of the highly resistant spores present in many animals that are capable of causing disabling disease in humans. This risk exists even for very low-level seepage via shallow surface drains or from rainfall run-off.

Some animal slurries are treated by mixing with straws and composted at high temperature to produce a bacteriologically sterile and friable manure for soil improvement.

Piggery Waste

Hog confinement barn or piggery

Piggery waste is comparable to other animal wastes and is processed as for general animal waste, except that many piggery wastes contain elevated levels of copper that can be toxic in the natural environment. The liquid fraction of the waste is frequently separated off and re-used in the piggery to avoid the prohibitively expensive costs of disposing of copper-rich liquid. Ascarid worms and their eggs are also common in piggery waste and can infect humans if wastewater treatment is ineffective.

Silage Liquor

Fresh or wilted grass or other green crops can be made into a semi-fermented product called silage which can be stored and used as winter forage for cattle and sheep. The production of silage often involves the use of an acid conditioner such as sulfuric acid or formic acid. The process of silage making frequently produces a yellow-brown strongly smelling liquid which is very rich in simple sugars, alcohol, short-chain organic acids and silage conditioner. This liquor is one of the most polluting organic substances known. The volume of silage liquor produced is generally in proportion to the moisture content of the ensiled material.

Treatment Silage liquor is best treated through prevention by wilting crops well before silage making. Any silage liquor that is produced can be used as part of the food for pigs. The most effective treatment is by containment in a slurry lagoon and by subsequent spreading on land following substantial dilution with slurry. Containment of silage liquor on its own can cause structural problems in concrete pits because of the acidic nature of silage liquor.

Milking Parlour (Dairy Farming) Wastes

Although milk has a deserved reputation as an important and valuable food product, its presence in wastewaters is highly polluting because of its organic strength, which can lead to very rapid de-oxygenation of receiving waters. Milking parlour wastes also contain large volumes of washdown water, some animal waste together with cleaning and disinfection chemicals.

Treatment Milking parlour wastes are often treated in admixture with human sewage in a local sewage treatment plant. This ensures that disinfectants and cleaning agents are sufficiently diluted and amenable to treatment. Running milking wastewaters into a farm slurry lagoon is a possible option although this tends to consume lagoon capacity very quickly. Land spreading is also a treatment option.

Slaughtering Waste

Wastewater from slaughtering activities is similar to milking parlour waste although considerably stronger in its organic composition and therefore potentially much more polluting.

Treatment As for milking parlour waste.

Vegetable Washing Water

Washing of vegetables produces large volumes of water contaminated by soil and vegetable pieces. Low levels of pesticides used to treat the vegetables may also be present together with moderate levels of disinfectants such as chlorine.

Treatment Most vegetable washing waters are extensively recycled with the solids removed by settlement and filtration. The recovered soil can be returned to the land.

Firewater

Although few farms plan for fires, fires are nevertheless more common on farms than on many other industrial premises. Stores of pesticides, herbicides, fuel oil for farm machinery and fertilizers can all help promote fire and can all be present in environmentally lethal quantities in firewater from fire fighting at farms.

Treatment All farm environmental management plans should allow for containment of substantial quantities of firewater and for its subsequent recovery and disposal by specialist disposal companies. The concentration and mixture of contaminants in firewater make them unsuited to any treatment method available on the farm. Even land spreading has produced severe taste and odour problems for downstream water supply companies in the past.

Compost

Compost is organic matter that has been decomposed and recycled as a fertilizer and soil amendment. Compost is a key ingredient in organic farming.

At the simplest level, the process of composting requires making a heap of wetted organic matter known as green waste (leaves, food waste) and waiting for the materials to break down into humus after a period of weeks or months. Modern, methodical composting is a multi-step, closely monitored process with measured inputs of water, air, and carbon- and nitrogen-rich materials. The decomposition process is aided by shredding the plant matter, adding water and ensuring proper aeration by regularly turning the mixture. Worms and fungi further break up the material. Bacteria requiring oxygen to function (aerobic bacteria) and fungi manage the chemical process by converting the inputs into heat, carbon dioxide and ammonium. The ammonium (NH_4) is the form of nitrogen used by plants. When available ammonium is not used by plants it is further converted by bacteria into nitrates (NO_3) through the process of nitrification.

A community-level composting plant in a rural area in Germany

Compost is rich in nutrients. It is used in gardens, landscaping, horticulture, and agriculture. The compost itself is beneficial for the land in many ways, including as a soil conditioner, a fertilizer, addition of vital humus or humic acids, and as a natural pesticide for soil. In ecosystems, compost is useful for erosion control, land and stream reclamation, wetland construction, and as landfill cover. Organic ingredients intended for composting can alternatively be used to generate biogas through anaerobic digestion.

Terminology

Composting of waste is an aerobic (in the presence of air) method of decomposing solid wastes. The process involves decomposition of organic waste into humus known as compost which is a good fertilizer for plants. However, the term "composting" is used worldwide with differing meanings. Some composting textbooks narrowly define composting as being an aerobic form of decomposition, primarily by aerobic or facultative microbes. An alternative form of organic decomposition to composting is "anaerobic digestion".

For many people, composting is used to refer to several different types of biological processes. In North America, "anaerobic composting" is still a common term for what much of the rest of the world and in technical publications people call "anaerobic digestion". The microbes used and the processes involved are quite different between composting and anaerobic digestion.

Ingredients

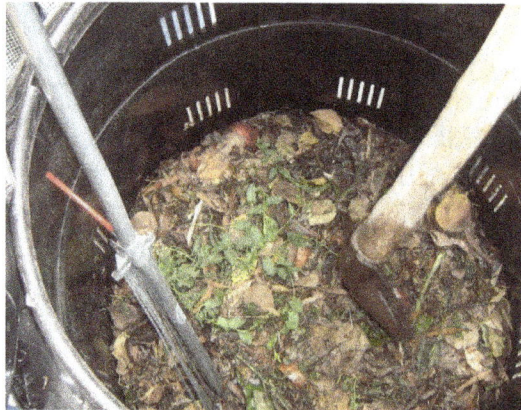

Home compost barrel in the Escuela Barreales, Santa Cruz, Chile

Carbon, Nitrogen, Oxygen, Water

Materials in a compost pile

Food scraps compost heap

Composting organisms require four equally important ingredients to work effectively:

- Carbon — for energy; the microbial oxidation of carbon produces the heat, if included at suggested levels.

    o   High carbon materials tend to be brown and dry.

- Nitrogen — to grow and reproduce more organisms to oxidize the carbon.

    o   High nitrogen materials tend to be green (or colorful, such as fruits and vegetables) and wet.

- Oxygen — for oxidizing the carbon, the decomposition process.

- Water — in the right amounts to maintain activity without causing anaerobic conditions.

Certain ratios of these materials will provide beneficial bacteria with the nutrients to work at a rate that will heat up the pile. In that process much water will be released as vapor ("steam"), and the oxygen will be quickly depleted, explaining the need to actively manage the pile. The hotter the pile gets, the more often added air and water is necessary; the air/water balance is critical to maintaining high temperatures (135°-160° Fahrenheit / 50° - 70° Celsius) until the materials are broken down. At the same time, too much air or water also slows the process, as does too much carbon (or too little nitrogen). Hot container composting focuses on retaining the heat to increase decomposition rate and produce compost quicker.

The most efficient composting occurs with an optimal carbon:nitrogen ratio of about 10:1 to 20:1. Rapid composting is favored by having a C/N ratio of ~30 or less. Theoretical analysis is confirmed by field tests that above 30 the substrate is nitrogen starved, below 15 it is likely to outgas a portion of nitrogen as ammonia. If nitrogen needs to be increased, it has been suggested to add 0.15 pounds of *actual* nitrogen per three bushels (3.75 cubic feet) of lower nitrogen material. [For those not familiar with these types of units: 0.64g/L or 640 grams of actual nitrogen per cubic meter.] Two to 3 pounds of organic nitrogen supplement (blood meal, manure, bone meal, alfalfa meal) per 100 pounds of low nitrogen materials (for example, straw or sawdust), supplies generally ample nitrogen and trace minerals in high carbon mixes.

Nearly all plant and animal materials have both carbon and nitrogen, but amounts vary widely, with characteristics noted above (dry/wet, brown/green). Fresh grass clippings have an average ratio of about 15:1 and dry autumn leaves about 50:1 depending on species. Mixing equal parts by volume approximates the ideal C:N range. Few individual situations will provide the ideal mix of materials at any point. Observation of amounts, and consideration of different materials as a pile is built over time, can quickly achieve a workable technique for the individual situation.

Animal Manure and Bedding

On many farms, the basic composting ingredients are animal manure generated on the farm and bedding. Straw and sawdust are common bedding materials. Non-traditional bedding materials are also used, including newspaper and chopped cardboard. The amount of manure composted on a livestock farm is often determined by cleaning schedules, land availability, and weather conditions. Each type of manure has its own physical, chemical, and biological characteristics. Cattle and horse manures, when mixed with bedding, possess good qualities for composting. Swine manure, which is very wet and usually not mixed with bedding material, must be mixed with straw or similar raw materials. Poultry manure also must be blended with carbonaceous materials - those low in nitrogen preferred, such as sawdust or straw.

Microorganisms

With the proper mixture of water, oxygen, carbon, and nitrogen, micro-organisms are allowed to break down organic matter to produce compost. The composting process is dependent on micro-organisms to break down organic matter into compost. There are many types of microorganisms found in active compost of which the most common are:

- Bacteria- The most numerous of all the microorganisms found in compost. Depending on the phase of composting, mesophilic or thermophilic bacteria may predominate.

- Actinobacteria- Necessary for breaking down paper products such as newspaper, bark, etc.

- Fungi- Molds and yeast help break down materials that bacteria cannot, especially lignin in woody material.

- Protozoa- Help consume bacteria, fungi and micro organic particulates.

- Rotifers- Rotifers help control populations of bacteria and small protozoans.

In addition, earthworms not only ingest partly composted material, but also continually re-create aeration and drainage tunnels as they move through the compost.

A lack of a healthy micro-organism community is the main reason why composting processes are slow in landfills with environmental factors such as lack of oxygen, nutrients or water being the cause of the depleted biological community.

Phases of Composting

Under ideal conditions, composting proceeds through three major phases:

- An initial, mesophilic phase, in which the decomposition is carried out under moderate temperatures by mesophilic microorganisms.

- As the temperature rises, a second, thermophilic phase starts, in which the decomposition is carried out by various thermophilic bacteria under high temperatures.

- As the supply of high-energy compounds dwindles, the temperature starts to decrease, and the mesophiles once again predominate in the maturation phase.

Human Waste

Human waste (excreta) can also be added as an input to the composting process, like it is done in composting toilets, as human waste is a nitrogen-rich organic material.

People excrete far more water-soluble plant nutrients (nitrogen, phosphorus, potassium) in urine than in feces. Human urine can be used directly as fertilizer or it can be put onto compost. Adding a healthy person's urine to compost usually will increase temperatures and therefore increase its ability to destroy pathogens and unwanted seeds. Urine from a person with no obvious symptoms of infection is much more sanitary than fresh feces. Unlike feces, urine does not attract disease-spreading flies (such as house flies or blow flies), and it does not contain the most hardy of pathogens, such as parasitic worm eggs. Urine usually does not stink for long, particularly when it is fresh, diluted, or put on sorbents.

Urine is primarily composed of water and urea. Although metabolites of urea are nitrogen fertilizers, it is easy to over-fertilize with urine, or to utilize urine containing pharmaceutical (or other) content, creating too much ammonia for plants to absorb, acidic conditions, or other phytotoxicity.

Humanure

"Humanure" is a portmanteau of *human* and *manure*, designating human excrement (feces and urine) that is recycled via composting for agricultural or other purposes. The term was first used in a 1994 book by Joseph Jenkins that advocates the use of this organic soil amendment. The term humanure is used by compost enthusiasts in the US but not generally elsewhere. Because the term "humanure" has no authoritative definition it is subject to various uses; news reporters occasionally fail to correctly distinguish between humanure and sewage sludge or "biosolids".

Uses

Compost is generally recommended as an additive to soil, or other matrices such as coir and peat, as a tilth improver, supplying humus and nutrients. It provides a rich *growing medium*, or a porous, absorbent material that holds moisture and soluble minerals, providing the support and nutrients in which plants can flourish, although it is rarely used alone, being primarily mixed with soil, sand, grit, bark chips, vermiculite, perlite, or clay granules to produce loam. Compost can be tilled directly into the soil or growing medium to boost the level of organic matter and the overall fertility of the soil. Compost that is ready to be used as an additive is dark brown or even black with an earthy smell.

Generally, direct seeding into a compost is not recommended due to the speed with which it may dry and the possible presence of phytotoxins that may inhibit germination, and the possible tie up of nitrogen by incompletely decomposed lignin. It is very common to see blends of 20–30% compost used for transplanting seedlings at cotyledon stage or later.

Composting can destroy pathogens or unwanted seeds. Unwanted living plants (or weeds) can be discouraged by covering with mulch/compost. The "microbial pesticides" in compost may include thermophiles and mesophiles, however certain composting detritivores such as black soldier fly larvae and redworms, also reduce many pathogens. Thermophilic (high-temperature) composting is well known to destroy many seeds and nearly all types of pathogens (exceptions may include prions). The sanitizing qualities of (thermophilic) composting are desirable where there is a high likelihood of pathogens, such as with manure.

Composting Technologies

A homemade compost tumbler

A modern compost bin constructed from plastics

Overview

In addition to the traditional compost pile, various approaches have been developed to handle different composting processes, ingredients, locations, and applications for the composted product.

There are a large number of different composting systems on the market, for example:

- At the household level: Composting toilet, container composting, vermicomposting

- At the industrial composting (large scale): Aerated Static Pile Composting, vermicomposting, windrow composting etc.

Examples

Vermicomposting

Rotary screen harvested worm castings

Food waste - after three years

Vermicompost is the product or process of composting through the utilization of various species of worms, usually red wigglers, white worms, and earthworms, to create a heterogeneous mixture of decomposing vegetable or food waste (excluding meat, dairy, fats, or oils), bedding materials, and vermicast. Vermicast, also known as worm castings, worm humus or worm manure, is the end-product of the breakdown of organic matter by species of earthworm. Vermicomposting is widely used in North America for on-site institutional processing of food waste, such as in hospitals and shopping malls. This type of composting is sometimes suggested as a feasible indoor home composting method. Vermicomposting has gained popularity in both these industrial and domestic settings because, as compared with conventional composting, it provides a way to compost organic materials more quickly (as defined by a higher rate of carbon-to-nitrogen ratio increase) and to attain products that have lower salinity levels that are therefore more beneficial to plant mediums.

The earthworm species (or composting worms) most often used are red wigglers (*Eisenia fetida* or *Eisenia andrei*), though European nightcrawlers (*Eisenia hortensis* or *Dendrobaena veneta*) could also be used. Red wigglers are recommended by most vermiculture experts, as they have some of the best appetites and breed very quickly. Users refer to European nightcrawlers by a variety of other names, including *dendrobaenas*, *dendras*, Dutch Nightcrawlers, and Belgian nightcrawlers.

Containing water-soluble nutrients, vermicompost is a nutrient-rich organic fertilizer and soil conditioner in a form that is relatively easy for plants to absorb. Worm castings are sometimes used as an organic fertilizer. Because the earthworms grind and uniformly mix minerals in simple forms, plants need only minimal effort to obtain them. The worms' digestive systems also add beneficial microbes to help create a "living" soil environment for plants.

Vermicompost tea in conjunction with 10% castings has been shown to cause up to a 1.7 times growth in plant mass over plants grown without.

Researchers from the Pondicherry University discovered that worm composts can also be used to clean up heavy metals. The researchers found substantial reductions in heavy metals when the worms were released into the garbage and they are effective at removing lead, zinc, cadmium, copper and manganese.

Hügelkultur (Raised Garden Beds or Mounds)

An almost completed Hügelkultur bed; the bed does not have soil on it yet.

The practice of making raised garden beds or mounds filled with rotting wood is also called "Hügelkultur" in German. It is in effect creating a Nurse log that is covered with soil.

Benefits of hügelkultur garden beds include water retention and warming of soil. Buried wood becomes like a sponge as it decomposes, able to capture water and store it for later use by crops planted on top of the hügelkultur bed.

The buried decomposing wood will also give off heat, as all compost does, for several years. These effects have been used by Sepp Holzer to enable fruit trees to survive at otherwise inhospitable temperatures and altitudes.

Black Soldier Fly Larvae Composting

Black Soldier Fly (*Hermetia illucens*) larvae have been shown to be able to rapidly consume large amounts of organic waste when kept at 31.8 °C, the optimum temperature for reproduction. Enthusiasts have experimented with a large number of different waste productsand some even sell starter kits to the public.

Cockroach Composting

Cockroach composting is another insect-mediated composting method. In this case the adults of any number of cockroach species (such as the Turkestan cockroach or *Blaptica dubia*) are used to quickly convert manure or kitchen waste to nutrient dense compost. Depending on species used and environmental conditions, excess composting insects can be used as an excellent animal feed for farm animals and pets.

Bokashi

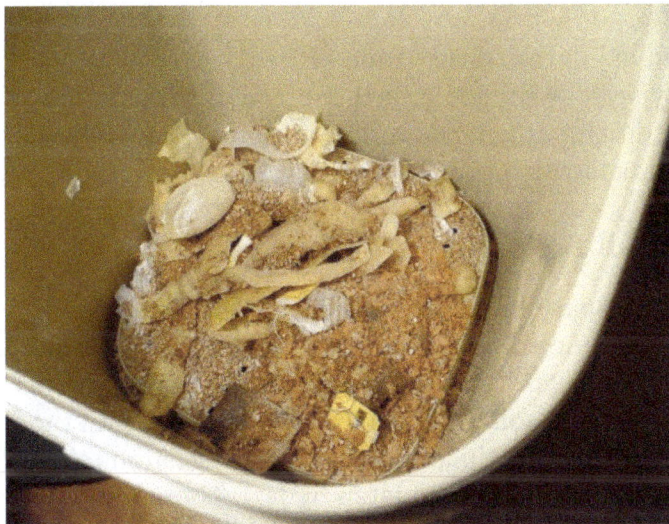

Inside a recently started bokashi bin. The aerated base is just visible through the food scraps and bokashi bran.

Bokashi is a method that uses a mix of microorganisms to cover food waste or wilted plants to decrease smell. Bokashi is Japanese for "shading off" or "gradation." It derives from the practice of Japanese farmers centuries ago of covering food waste with rich, local soil that contained the microorganisms that would ferment the waste. After a few weeks, they would bury the waste.

Most practitioners obtain the microorganisms from the product Effective Microorganisms (EM1), first sold in the 1980s. EM1 is mixed with a carbon base (e.g. sawdust or bran) that it sticks to and a sugar for food (e.g. molasses). The mixture is layered with waste in a sealed container and after a few weeks, removed and buried.

Newspaper fermented in a lactobacillus culture can be substituted for bokashi bran for a successful bokashi bucket.

Compost Tea

Compost teas are defined as water extracts brewed from composted materials and can be derived from aerobic or anaerobic processes. Compost teas are generally produced from adding one volume of compost to 4-10 volumes of water, but there has also been debate about the benefits of aerating the mixture. Field studies have shown the benefits of adding compost teas to crops due to the adding of organic matter, increased nutrient availability and increased microbial activity. They have also been shown to have an effect on plant pathogens.

Composting Toilets

A composting toilet does not require water or electricity, and when properly managed does not smell. A composting toilet collects human excreta which is then added to a compost heap together with sawdust and straw or other carbon rich materials, where pathogens are destroyed to some extent. The amount of pathogen destruction depends on the temperature (mesophilic or thermophilic conditions) and composting time. A composting toilet tries to process the excreta in situ although this is often coupled with a secondary external composting step. The resulting compost product has been given various names, such as humanure and EcoHumus.

A composting toilet can aid in the conservation of fresh water by avoiding the usage of potable water required by the typical flush toilet. It further prevents the pollution of ground water by controlling the fecal matter decomposition before entering the system. When properly managed, there should be no ground contamination from leachate.

Compost and Land-filling

As concern about landfill space increases, worldwide interest in recycling by means of composting is growing, since composting is a process for converting decomposable organic materials into useful stable products. Composting is one of the only ways to revitalize soil vitality due to phosphorus depletion in soil. Industrial scale composting in the form of in-vessel composting, aerated static pile composting, and anaerobic digestion takes place in most Western countries now, and in many areas is mandated by law. There are process and product guidelines in Europe that date to the early 1980s (Germany, the Netherlands, Switzerland) and only more recently in the UK and the US. In both these countries, private trade associations within the industry have established loose standards, some say as a stop-gap measure to discourage independent government agencies from establishing tougher consumer-friendly standards. The USA is the only Western country that does not distinguish sludge-source compost from green-composts, and by default in the USA 50% of states expect composts to comply in some manner with the federal EPA 503 rule promulgated in 1984 for sludge products. Compost is regulated in Canada and Australia as well.

Industrial Systems

A large compost pile that is steaming with the heat generated by thermophilic microorganisms.

Industrial composting systems are increasingly being installed as a waste management alternative to landfills, along with other advanced waste processing systems. Mechanical sorting of mixed waste streams combined with anaerobic digestion or in-vessel composting is called mechanical biological treatment, and is increasingly being used in developed countries due to regulations controlling the amount of organic matter allowed in landfills. Treating biodegradable waste before it enters a landfill reduces global warming from fugitive methane; untreated waste breaks down anaerobically in a landfill, producing landfill gas that contains methane, a potent greenhouse gas.

Vermicomposting, also known as vermiculture, is used for medium-scale on-site institutional composting, such as for food waste from universities and shopping malls. It is selected either as a more environmentally friendly choice than conventional methods of disposal, or to reduce the cost of commercial waste removal.

Large-scale composting systems are used by many urban areas around the world. Co-composting is a technique that combines solid waste with de-watered biosolids, although difficulties controlling inert and plastics contamination from municipal solid waste makes this approach less attractive. The world's largest MSW co-composter is the Edmonton Composting Facility in Edmonton, Alberta, Canada, which turns 220,000 tonnes of residential solid waste and 22,500 dry tonnes of biosolids per year into 80,000 tonnes of compost. The facility is 38,690 m² (416,500 sq.ft.) in area, equivalent to 4½ Canadian football fields, and the operating structure is the largest stainless steel building in North America, the size of 14 NHL rinks. In 2006, Qatar awarded Keppel Seghers Singapore, a subsidiary of Keppel Corporation, a contract to begin construction on a 275,000 tonne/year anaerobic digestion and composting plant licensed by Kompogas (de) Switzerland. This plant, with 15 independent anaerobic digesters, will be the world's largest composting facility once fully operational in early 2011 and forms part of Qatar's Domestic Solid Waste Management Centre, the largest integrated waste management complex in the Middle East.

Another large MSW composter is the Lahore Composting Facility in Lahore, Pakistan, which has a capacity to convert 1,000 tonnes of municipal solid waste per day into compost. It also has a capacity to convert substantial portion of the intake into refuse-derived fuel (RDF) materials for further combustion use in several energy consuming industries across Pakistan, for example in cement manufacturing companies where it is used to heat cement kilns. This project has also been approved by the Executive Board of the United Nations Framework Convention on Climate Change for reducing methane emissions, and has been registered with a capacity of reducing 108,686 tonnes carbon dioxide equivalent per annum.

Related Technologies

Anaerobic digestion is process for converting organic waste into (biogas). The residual material, sometimes in combination with sewage sludge can be followed by an aerobic composting process before selling or giving away the compost.

History

Composting as a recognized practice dates to at least the early Roman Empire since Pliny the Elder (AD 23-79). Traditionally, composting involved piling organic materials until the next planting season, at which time the materials would have decayed enough to be ready for use in the soil. The advantage of this method is that little working time or effort is required from the composter and it fits in naturally with agricultural practices in temperate climates. Disadvantages (from the modern perspective) are that space is used for a whole year, some nutrients might be leached due to exposure to rainfall, and disease-producing organisms and insects may not be adequately controlled.

Composting was somewhat modernized beginning in the 1920s in Europe as a tool for organic farming. The first industrial station for the transformation of urban organic materials into compost was set up in Wels, Austria in the year 1921. Early frequent citations for propounding composting within farming are for the German-speaking world Rudolf Steiner, founder of a farming method called biodynamics, and Annie Francé-Harrar, who was appointed on behalf of the government in Mexico and supported the country 1950–1958 to set up a large humus organization in the fight against erosion and soil degradation.

Compost Basket

In the English-speaking world it was Sir Albert Howard who worked extensively in India on sustainable practices and Lady Eve Balfour who was a huge proponent of composting. Composting was imported to America by various followers of these early European movements by the likes of J.I. Rodale (founder of Rodale Organic Gardening), E.E. Pfeiffer (who developed scientific practices in biodynamic farming), Paul Keene (founder of Walnut Acres in Pennsylvania), and Scott and Helen Nearing (who inspired the back-to-the-land movement of the 1960s). Coincidentally, some of the above met briefly in India - all were quite influential in the U.S. from the 1960s into the 1980s.

There are many modern proponents of rapid composting that attempt to correct some of the perceived problems associated with traditional, slow composting. Many advocate that compost can be made in 2 to 3 weeks. Many such short processes involve a few changes to traditional methods, including smaller, more homogenized pieces in the compost, controlling carbon-to-nitrogen ratio (C:N) at 30 to 1 or less, and monitoring the moisture level more carefully. However, none of these parameters differ significantly from the early writings of Howard and Balfour, suggesting that in fact modern composting has not made significant advances over the traditional methods that take a few months to work. For this reason and others, many modern scientists who deal with carbon transformations are sceptical that there is a "super-charged" way to get nature to make compost rapidly.

In fact, both sides are right to some extent. The bacterial activity in rapid high heat methods breaks down the material to the extent that pathogens and seeds are destroyed, and the original feedstock is unrecognizable. At this stage, the compost can be used to prepare fields or other planting areas. However, most professionals recommend that the compost be given time to cure before using in a nursery for starting seeds or growing young plants. The curing time allows fungi to continue the decomposition process and eliminating phytotoxic substances.

Many countries such as Wales and some individual cities such as Seattle and San Francisco require food and yard waste to be sorted for composting.

Kew Gardens in London has one of the biggest non-commercial compost heaps in Europe.

Composting

Biodegradation is a natural, ongoing biological process that is a common occurrence in both human-made and natural environments. In the broadest sense, any organic material that can be biologically decomposed is compostable. In fact, human beings have used this naturally occurring process for centuries to stabilise and recycle agricultural and human wastes.Today, however,composting is a diverse practice that includes a variety of approaches, depending on the type of organic materials being composted and the desired properties of the final product.

To derive the maximum benefit from the natural, but typically slow decomposition process (e.g., grass clippings left in the lawn or food scraps rotting in dustbins, etc.), it is necessary to control the environmental conditions during the composting process. The overall composting process can be explained as follows:

Organic matter + O_2 + aerobic bacteria $\rightarrow$ CO_2 + NH_3 + H_2O + other end products+ energy

Compost is the end product of the composting process. The by-products of this process are carbon dioxide and water. Compost is peaty humus, dark in colour and has a crumbly texture, an earthy odour, and resembles rich topsoil. Composts will not have any resemblance in the physical form to the original waste from which it was derived. High-quality compost is devoid of weed seeds and organisms that may be pathogenic to humans, animals, or plants. Cured compost is also relatively stable and resistant to further decomposition by microorganisms.

A complete composting process is shown in Figure:

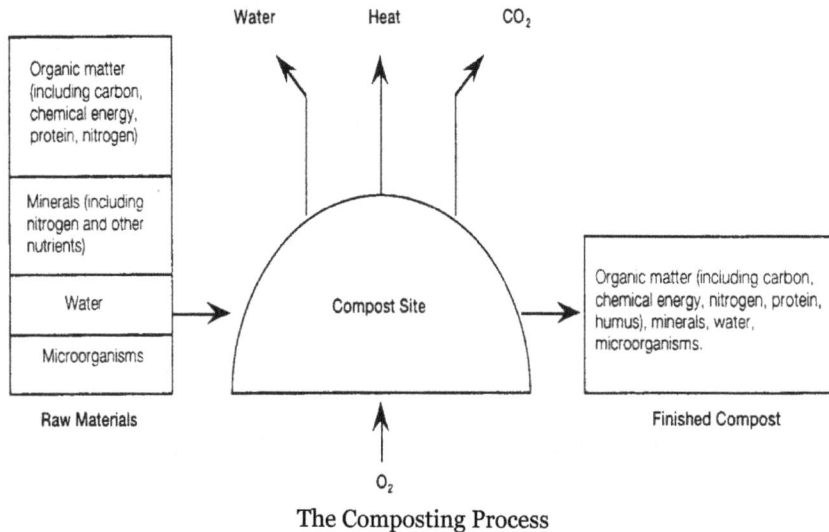

The Composting Process

As mentioned earlier, the composting process is an environmentally sound and beneficial means of recycling organic materials and not a means of waste disposal. And, it is important to view compostable materials as usable and not as waste requiring disposal. Before we discuss the composting process in detail, let us talk about the benefits of composting.

Benefits

Composting is one of the important components of solid waste management (SWM). It is a form of source reduction or waste prevention, as the materials are completely diverted from the disposal facilities and require no management or transportation. Community-yard trimming composting programme, source- separated organic composting and mixed municipal solid waste (MSW) composting constitute the various recycling processes. A major portion of municipal solid wastes in India contain up to 70% by weight of organic materials.

In addition, certain industrial by-products – thosefromfoodprocessing, agricultural and paper industries – are mostly composed of organic materials.

Composting, being an organic material, can significantly reduce waste stream volume. Diverting such materials from the waste stream frees up landfill space needed for materials that cannot be composted. Composting owes its current popularity to several factors, including increased landfill tipping fees, shortage of landfill capacity and increasingly restrictive measures imposed by regulatory agencies. In addition, recycling mandates indirectly encourage composting, as they consider it an acceptable strategy for achieving mandatory goals.

Composting may also offer an attractive economic advantage for communities where the costsof using other options are high. However, it is considered a viable option only when the compost can be marketed. In some cases, nevertheless, the benefits of reducing disposal needs through composting may be adequate to justify choosing this option, even if the compost is only used as a landfill cover. Compost, because of its high organic matter content, makes a valuable soil amendment and is used to provide nutrients for plants. When mixed with soil, compost promotes a proper balance between air and water inthe resulting mixture, helps reduce soil erosion and serves as a slow-release fertiliser.

Processes

Several factors contribute to the success of composting, including physical, chemical, and biological processes (EPA, 1989 and 1995). Understanding these processes, therefore, is necessary for making informed decisions, when developing and operating a composting programme.

Biological Processes

Microorganisms such as bacteria, fungi and actinomycetes as well as larger organisms such as insects and earthworms play an active role in decomposing the organic materials. As microorganisms begin to decompose the organic material, they break down organicmatter and produce carbon dioxide, water,heat and humus (the relatively stable organic end product). This humus end product is compost.

Microorganisms consume some of the carbon to form new microbial cells, as they increase their population. They have, however, preferences on the type of organic material they consume. When the organic molecules they require are not available, they may become dormant or die. The chain of succession of different types of microbes continues, until there is little decomposable organic material left. At this point, the organic material that is remaining is termed compost. It is largely made up of microbial cells, microbial skeletons and by-products of microbial decomposition, and undecomposed particles of organic and inorganic origin. Decomposition may proceed slowly at first because of smaller microbial populations, but as populations grow in the first few hours or days, they rapidly consume the organic materials present in the feedstock. The number and kind of microorganisms are generally not a limiting environmental factor in composting non-toxic agricultural materials, yard trimmings, or municipal solid wastes, all of which usually contain an adequate diversity of microorganisms.

(i) Mesophilic, or moderate-temperature phase: Compost bacteria combine carbon with oxygen to produce carbon dioxide and energy. The microorganisms for reproduction and growth use some of the energy and the rest is generated as heat. When a pile of organic refuse begins to undergo the composting process, mesophilic bacteria proliferate, raising the temperature of the composting mass up to 44°C. This is the first stage of the composting process. These mesophilic bacteria can include E. coli and other bacteria from the human intestinal tract, but these soon become increasingly inhibited by the temperature, as the thermophilic bacteria take over in the transition range of 44°C – 52°C.

(ii) Thermophilic, or high-temperature phase: In the second stage of the process, the thermophilic microorganisms are very activeandproduce heat. This stage can continue up to about 70°C, although such high temperatures are neither common nor desirable in compost. This heating stage takes place rather quickly and may last only a few days, weeks, or months. It tends to remain localised in the upper portion of a compost pile where the fresh material is being added, whereas in batch compost, the entire composting mass may be thermophilicall at once. After the thermophilic heating period, the manure will appear to have been digested, but the coarser organic material will not be digested. This is when the third stage of composting, i.e., the cooling phase, takes place.

(iii) Cooling phase: During this phase, the microorganisms that were replaced by the thermophiles migrate back into the compost and digest the more resistant organic materials. Fungi andmacroorganismssuchas earthworms and sow bugs that break the coarser elementsdown into humus also move back in.

(iv) Maturation or curing phase: The final stage of the composting process is called curing, ageing, or maturing stage, and is a long and important one. A long curing period (e.g., a year after the thermophilic stage) adds a safety net for pathogen destruction. Manypathogens have a limited period of viability in the soil, and the longer they are subjected to the microbiological competition of the compost pile the more likely they will die a swift death. Immature compost can be harmful to plants. Uncured compost can, for example, produce phytotoxins (i.e., substances toxic to plants), robbing the soil of oxygen and nitrogen and contain high levels of organic acids.

Different communities of microorganisms predominate during the various composting phases. Initial decomposition is carried out by mesophilic microorganisms, which rapidly break down the soluble, readily degradable compounds. The heat they produce causes the compost temperature to rise rapidly. As the temperature rises above 40°C, the mesophilic microorganisms become less competitive and are replaced by thermophilic (heat loving) ones. Attemperatures of 55°C and above, many microorganisms that are pathogenic to humans or plants are destroyed. Temperatures above 65°C kill many forms of microbes and limit the rate of decomposition. Compost managers use aeration and mixing to keep the temperature below this point. During the thermophilic phase, high temperatures accelerate the breakdown of proteins, fats, and complex carbohydrates like cellulose and hemicellulose, the major structural molecules in plants. As the supply of these high-energy compounds become exhausted, the compost temperature gradually decreases and mesophilic microorganisms once again take over the final phase of curing or maturation of the remaining organic matter.

The composting process, therefore, should cater to the needs of the microorganisms and promote conditions that will lead to rapid stabilisation of the organic materials. While several microorganisms are beneficial to the composting process and may be present in the final product, some microbes are potential pathogens to animals, plants or humans. These pathogenic organisms must be destroyed in the composting process before the compost is distributed in the market place. Most of this destruction takes place by controlling the temperature of composting operations.

Chemical Processes

Several factors determine the chemical environment for composting. These include the presence of an adequate carbon food/energy source, a balanced amount of nutrients, the correct amount of water, adequate oxygen, appropriate pH and the absence of toxic constituents that couldinhibitmicrobialactivity (EPA, 1989 and 1995). Let us now describe each of these factors below:

(i) Carbon/energy source: For their carbon/energy source, micro organisms in the composting process rely on carbon in the organic material, unlike higher plantsthat rely on carbon dioxide and sunlight. The carbon contained in natural or human-based organic materials may or may not be biodegradable.The relative ease with which a material is biodegraded depends on the genetic makeup of the micro organisms present and the organic molecules that the organism decomposes.

Since most municipal and agricultural organics and yard trimmings contain an adequate amount of biodegradable forms of carbon, it is not a limiting factor in the composting process. As more easily degradable forms of carbon are decomposed, a small portion of the carbon is converted into microbial cells, and a significant portion is converted to carbon dioxide and lost to the atmosphere. As the composting process progresses, the loss of carbon results in a decrease in weight and volume of the feedstock. The less-easily decomposed forms of carbon will form the matrix for the physical structure of the final product.

(ii) Nutrients: Among the plant nutrients (i.e., nitrogen, phosphorus, and potassium), nitrogen is of greatest concern, because it is lacking in some plant materials. The carbon-nitrogen ratio, which is established on the basis of available carbon rather than total carbon, is considered critical in determining the rate of decomposition. Leaves, for example, are a good source for carbon, and fresh grass, manure and slaughterhouse waste are the sources for nitrogen. To aid the decomposition process, the bulk of the organic matter should be carbon with just enough nitrogen. In general, an initial ratio of 30:1 (C: N or Carbon: Nitrogen) is considered ideal. Higher ratios tend to retard the process of decomposition, while ratios below 25:1 may result in odour problems. As the composting process proceeds and carbon is lost to the atmosphere, this ratio decreases.

Finished compost should have ratios of 15 to 20:1. Adding 3 – 4 kg of nitrogen material for every 100 kg of carbon should be satisfactory for efficient and rapid composting. To lower the carbon to nitrogen ratios, nitrogen-rich materials such as yard trimmings, animal manures, or bio- solids are often added. Adding partially decomposed orcomposted materials (with a lower carbon: nitrogen ratio) as inoculums may also lower the ratio. As the temperature in the compost pile rises and carbon: nitrogen ratio falls below 25:1, the nitrogen in the fertiliser is lost as gas (ammonia)to the atmosphere. The composting process slows, if there is not enough nitrogen, and too much nitrogen may cause the generation of ammonia gas, which can create unpleasant odours.

(iii) Moisture:Waterisanessentialpartofallformsoflife,andthemicroorganisms living in a compost pile are no exception. Since most compostable materials have lower than ideal water content, i.e., 50 to 60% of total weight, the composting process may be slower than desired,if water is not added. However, it should not be high enough to create excessive free flow of water and movement caused by gravity. Excessive moisture and flowing water form leachate, which creates potential liquid management problems including water and air pollution (e.g., odour). For example, excess moisture impedes oxygen transfer to the microbial cells, can increase the possibility of developing anaerobic conditions and may lead to rotting and obnoxious odours. Microbial processes contribute moisture to the compost pile during decomposition. Although moisture is added, it is also being lost through evaporation. Since the amount of water evaporated usually exceeds the input of moisture from the decomposition processes, there is generally a net loss of moisture from the compost pile. In such cases, adding moisture may be necessary to keep the composting process performing at its peak.

Controlling the size of piles can minimise evaporation from compost piles, as piles with larger volumes have less evaporating surface/unit volume than smaller piles. The water added must be thoroughly mixed so thatthe organic fraction in the bulk of the material is uniformly wetted and composted under ideal conditions. Properly wetted compost has the consistency of a wet sponge. Systems that facilitate the uniform addition of water at any point in the composting process are preferable.

(iv) Oxygen: Composting is considered an aerobic process. Decomposition can occur under both aerobic (requiring oxygen) and anaerobic (lacking oxygen) conditions. The compost pile should have enough void space to allow free air movement so that oxygen from the atmosphere can enter the pile and the carbon dioxide and other gases emitted can be exhausted tothe atmosphere. To maintain aerobic conditions, in which decomposition occurs at a fast rate, the compost pile is mechanically aerated or turned frequently to expose the microbes to the atmosphere and to create more air spaces by fluffing up the pile.

A 10 to 15% oxygen concentration is considered adequate, although a concentration as low as 5% may be sufficient for leaves. While higher concentrations of oxygen will not negatively affect the composting process, circulation of an excessive amount of air can cause problems. For example, excess air removes heat, which cools the pile and also promotes excess evaporation. In other words, excess air slows down the rate of composting. Excess aeration is also an added expense that increases production costs.

(v) pH: The pH factor affects the amount of nutrients available for the microorganisms, the solubility of heavy metals and the overall metabolic activity of the microorganisms. A pH between 6 and 8 is considered optimum, and it can be adjusted upward by the addition of lime, or downward with sulphur, although such additions are normallynot necessary. The composting process itself produces carbon dioxide, which, when combined with water, produces carbonic acid, which could lower the pH of the compost. As the composting process progresses, the final pH varies, depending on the specific type of feedstock used and operating conditions. Wide swings in pH are unusual, since organic materials are naturally well buffered with respect to pH changes. Note that down swings in pH during composting usually do not occur.

What the foregoing discussion informs us is that the composition of material to be composted largely determines the chemical environment. In addition, several modifications can be made during the composting process to create an ideal chemical environment for the rapid decomposition of organic materials.

The physical environment in the compost process includes factors such as temperature, particle size, mixing and pile size (EPA, 1989 and 1995). Each of these is essential for the composting process to proceed in an efficient manner as described below:

(i) Particle size: As composting progresses, there is a natural process of size reduction and the particle size of the material being composted is critical. Because smaller particles usually have more surface area per unit weight, they facilitate more microbial activity on their surfaces, which leads to rapid decomposition. The optimum particle size has enoughsurfaceareafor rapid microbial activity and also enough void space to allow air to circulate for microbial respiration. The feedstock composition can be manipulated to create the desired mix of particle size and void space. For instance, through particle size reduction, we can increase the desired combination of void space and surface area for garden trimmings or municipal solid wastes. To improve the aesthetic appeal of finished composts, we can carryout particle size reduction, after the composting process is completed.

(ii) Temperature: Composting can occur at a range of temperatures, and the optimum temperature range is between 32° and 60° C.Temperatures above 65° C are not ideal for composting as thermal destruction of cell proteins kill the organisms. Similarly, temperatures below the minimum required for a group of organisms affect the metabolic activity (i.e., regulatory machinery) of the cells. Temperatures can be lowered, if required, by either increasing the frequency ofmechanicalagitationor using airflow throttling, temperature feedback control or blowers controlled with timers. Mixing or mechanical aeration also provides air for the microbes.

When compost is at a temperature greater than 55° C for at least three days, pathogen destruction occurs. It is important that all portions of the compost material are exposed to such temperaturesto ensure pathogendestruction throughout the compost. At these temperatures, weed seeds are also destroyed. After the complete pathogen destruction, temperatures may be lowered and maintained at slightly lower levels (51° to 55° C). Attaining and maintaining 55° C for three days is not difficult for in-vessel composting systems. However, to achieve pathogen destruction with windrow composting systems, the 55° C temperature level must be maintained for a minimum of 15 days, during which time the windrows must be turned at least five times. The longer duration and increased turning are necessary to achieve a uniform pathogen destruction in the entire pile. Care should be takentoavoid contact between materials that have achieved these minimum temperatures and materials that have not.

(iii) Mixing: Mixing of feedstock, water and inoculants is important and is done by turning or mixing the piles after composting has begun. Mixing and agitation distribute moisture and air evenly, and promote the breakdown of compost clumps. Excessive agitation of open vessels or piles, however, can cool them and retard microbial activity.

Stages

There are five basic stages involved in all composting practices, namely preparation, digestion, curing, screening or finishing, and storage or disposal. However, you must note that differences (among various composting processes) may occur in the method of digestion or in the amount of preparation and the finishing required. In choosing the type of process to be used and the amount of sophistication required, a number of criteria must be considered (Pavoni, et al., 1975).

Preparation

This preparation phase of composting involves several steps, and these depend upon the sophistication of the plant and the amount of resource recovery practised. A typical preparation process, however, may include such activities as the sorting of recyclable materials, the removal of non-combustibles, the shredding, pulping, grinding and the adding of water sludge. Most plants utilise receiving equipment, which provides a steady flow of solid waste throughout the operation. Consistency of flow is accomplished by the use of storage hoppers and regulated conveyor system. After the solid wastes leave the receiving area, the bulky items, which could damage the grinders, are removed by hand. The separation of other non-compostable recyclable materials like glass, metal, rag, plastic, rubber and paper may be accomplished before or after comminution (i.e., reduction to small pieces or particles by pounding or abrading) by either hand or mechanical means. Those non-compostables, which are not salvaged, must be ultimately disposed of in a sanitary landfill.

After initial separation, most composting processes require the solid waste material to be reduced in particle size to facilitate handling, digestion and mixing of the product to provide more homogenous compost. The three major methods of comminution, which have been utilised in composting processes, include hammermill,rasper/shredderandwetpulper,andthevariousequipmentrequired are scales, receiving bins, conveyors, grinders and screens.

Since the refuse characteristics vary from one load to the next, a final step in the preparation phase of composting may be to adjust the moisture and nitrogen content of the solid waste to be composted. The optimummoisture content ranges from 45 to 55% of wet weight, while the optimum carbon to nitrogen ratio should be below 30%. The moisture and nutrient adjustments can be accomplished most efficiently through the addition of raw wastewater sludge. This increases the volume of composted material by 6 to 10%, in addition to accelerating the composting operation and producing a better final produc t in terms of nutrient contents.

Digestion

Digestion techniques are the most unique feature of the various composting processes and may vary from the backyard composting process to the highly controlled mechanical digester. Composting systems fall into the following two categories:

(i) windrow composting in open windrows;

(ii) mechanical composting in enclosed digestion chambers.

Curing

Organic materials, remaining after the first (rapid) phase of composting, decompose slowly, despite ideal environmental conditions. The second phase, which is usually carried out in windrows, typically takes from a few weeks to six months, depending on the outdoor temperatures, intensity of management and market specifications for maturity. With some system configurations, a screening step may precede the curing operation. During curing, the compost becomes biologically stable, with microbial activity occurring at a slower rate than that during actual composting. Curing piles may be either force-aerated or passive aerated with occasional turning. As the pile cures, the microorganisms generate less heat and the pile begins to cool.

Note that the cooling of piles does not always mean that the curing is complete. Cooling is merely a sign of reduced microbial activity, which can result from lack of moisture, inadequate oxygen within the pile, nutrient imbalance or the completion of the composting process. Curing may take from a few days to several months to complete. The cured compost is then marketed.

Screening or Finishing

Compost is screened or finished to meet the market specifications. Sometimes, this processing is done before the compost is cured. One or two screening steps and additional grindings are used to prepare the compost for markets. During the composting operation, the compostable fraction separated from the non- compostable fraction, through screens, undergoes a significant size reduction. The non-compostable fraction retained on the coarse screen issenttothe landfill, while the compostable materials retained on finer screens may be returned to the beginning of the composting process to allow further composting. The screened compost may contain inert particles such as glass or plastic that may have passed through the screen. The amount of such inert materials depends on feedstock processing before composting and the composting technology used.

To successfully remove the foreign matter and recover the maximum compost by screening, the moisture content should be below 50%. Drying should be allowed only after the compost has sufficiently cured. If screening takes place before curing is complete, moisture addition may be necessary to cure the compost. The screen size used is determined by market specifications of particle size.

Storage or Disposal

In the final analysis, regardless of the efficiency of the composting process, the success or failure of the operation depends upon the method of disposal. Even when a good market for compost exists, provision must be still made for storage. Storage is necessary because the use of composting is seasonal, with greatest demand during spring and winter. Therefore, a composting plant must have a 6- month storage area. For a 300 tonne per day plant, this will require about 6 hectares of storage area. Many composting operations combinetheircuring period with the storage period. The price at which compost can be sold depends on the benefit to be obtained from its use and the customers who are willing to make use of such benefits. Compost may be sold in bulk, finished or unfinished, as well as fortified with chemical fertilisers.

Technologies

The composting technologies – windrow, aerated static pile, in-vessel composting and anaerobic processing (EPA, 1989 and 1995) – vary in the method of air supply, temperature control, mixing/turning of the material, time required for composting, and capital and operating costs. Besides these general categories of composting technologies, there are also some supporting technologies, which include sorting, screening, and curing.

Let us discuss the four general categories of composting technologies, next.

Windrow Composting

The windrow system is the least expensive and most common approach. Windrows are defined as regularly turned elongated piles, shaped like a haystack in cross section and up to a hundred meters or more in length. The cross - sectional dimensions vary with feedstock and turning equipment, but most municipal solid waste (MSW) windrows are 1.5 to 3 meters high and 3 to 6 meters wide as shown in Figure below:

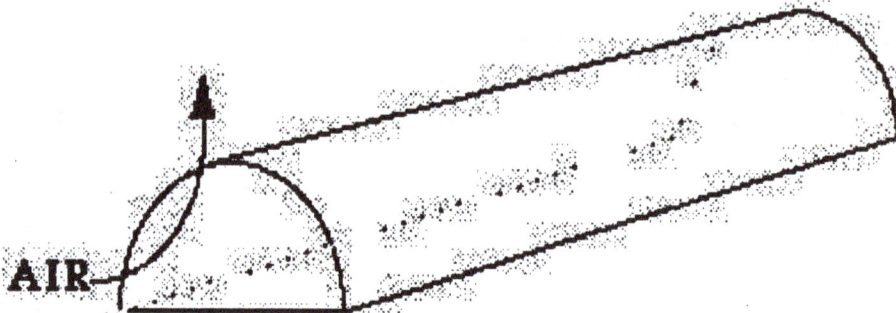

Windrow Composting

Windrows composed of MSW are usually required to be located on an impermeable surface, which greatly improves equipment handlingunder inclement weather conditions. The optimum size and shape of the windrow depends on particle size, moisture content, pore space and decomposition rate – all of which affect the movement of oxygen towards the centre of the pile.

Process control is normally through pile management, although forced aeration can also be used. Turning the pile re-introduces air into the pile and increases porosity so that efficient passive aeration from atmospheric air continues at all times. The windrow dimensions should allow conservation of the heat generated during the composting process and also allow air to diffuse to the deeper portions of the pile.

As mentioned earlier, windrows must be placed on a firm surface to turn the piles with ease. They may be turned as frequently as once per week, but more frequent turning may be necessary, if high proportions of bio-solids are present in the feedstock. Turning the piles also moves the materials from the pile surface to the core of the windrow, where they can undergo composting. Machines equipped with augers, paddlesor tinesare used for turning the piles. Some windrow turners can supplement piles with water, if necessary. When piles are turned, heat is released as steam to the atmosphere. If inner portions of the pile have low levels of oxygen, odours may result when this portion of the pile is exposed to the atmosphere. Piles with initial moisture content within the optimum range have a reduced potential for producing leachate. The addition of moisture from precipitation, however, increases this potential.

Any leachate or runoff created must be collected and treated or added to a batch of incoming feedstock to increase its moisture content. To avoid problems with leachate or runoff, piles can be placed under a roof, but doing so adds to the initial costs of the operation.

Aerated Static Pile Composting

Aerated static pile composting is a non-proprietary technology that requires the composting mixture (i.e., a mixture of pre-processed materials and liquids) to be placed in piles that are mechanically aerated. The piles are placed overa network of pipes connected to a blower, as in Figure, which supplies the air for composting:

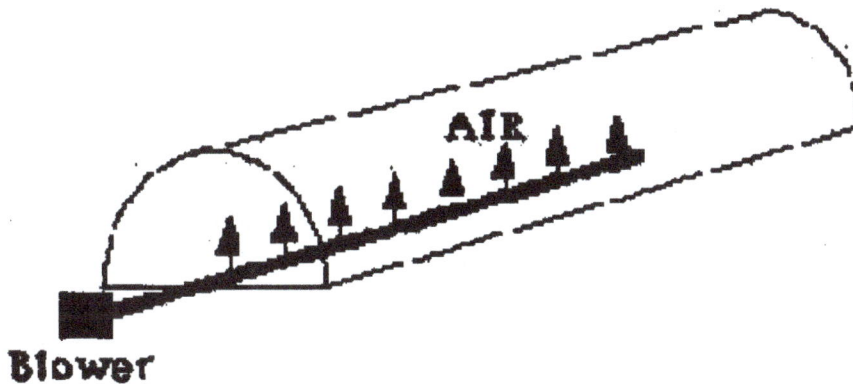

Aerated Static Pile Composting

Air can be supplied under positive or negative pressure. That is to say, the air supply blower either forces air into the pile or draws air out of it. Forcing the air into the pile generates a positive pressure system, while drawing air out of the pile creates a negative pressure. When the composting process is nearly complete, the piles are broken up for the first time since their construction. The compost is then taken through a series of post-processing steps. A timer or a temperature feedback system similar to a home thermostat controls the blowers.

Air circulation in the compost piles provides the needed oxygen for the composting microbes and prevents excessive heat build-up in the pile. Removing excess heat and water vapour cools the pile to maintain optimum temperature for microbial activity. A controlled air supply enables construction oflargepiles, which decreases the need for land. Odours from the exhaust air could be substantial, but traps or filters can be used to control them. The temperatures in the inner portion of a pile are usually adequate to destroy a significant number of the pathogens and weed seeds present. The surface of piles, however, may not reach the desired temperatures for destruction of pathogens because piles are not turned in the aerated static pile technology. This problem can be overcome by placing a layer of finished compost of 15 to 30 cms thick over the compost pile. The outer layer of finished compost acts as an insulating blanket and helps maintain the desired temperature for destruction of pathogens and weed seeds throughout the entire pile.

Aerated static pile composting systems have been used successfully for MSW, yard trimmings, bio-solids and industrial composting. Aerated staticpile composting can also be done under a roof or in the open. Producing compost using this technology usually takes about 6 to 12 weeks. The land requirements for this method are lower than that of windrow composting.

In-vessel Composting System

In-vessel composting systems enclose the feedstock in a chamber or vessel that provides adequate mixing, aeration and moisture. Drums, digester binsand tunnels are some of the common in-vessel type systems. In-vessel systems vary in their requirements for pre-processing materials. For example, some require minimal pre-processing, while others require extensive MSW pre-processing. These vessels can be single- or multi-compartment units. In some cases, the vessel rotates, and in others, it is stationary and a mixing/agitating mechanism moves thematerial around. Most in-vessel systems arecontinuous-feedsystems, although some operate in a batch mode. All in-vessel systems require further composting (curing) after the material has been discharged fromthe vessel.

A major advantage of in-vessel systems is that all environmental conditions can be carefully controlled to allow rapid composting. The material to be composted is frequently turned and mixed to homogenise the compost and promote rapid oxygen transfer. Retention times range from less than one week to as long as four weeks. These systems, if properly operated, produce minimal odours and little or no leachate. In-vessel systems enable exhaust gases from the vessel to be captured and are subjected to odour control and treatment. Some of the commonly used in-vessel systems are as follows:

- Vertical composting reactor: It is generally over 4 meters high as illustrated in Figure, and can be housed in silos or other large structures:

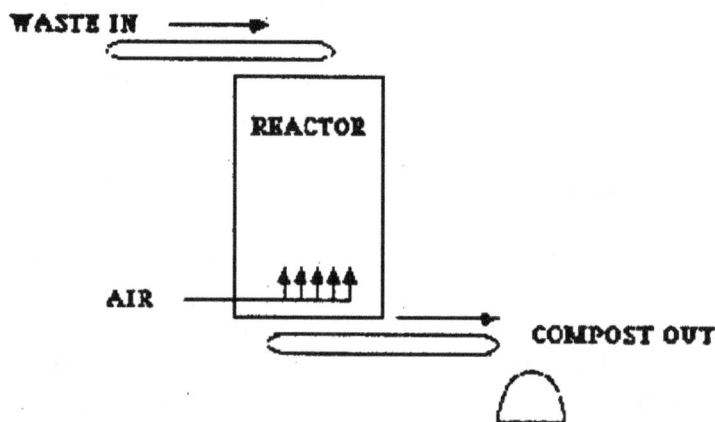

Vertical Composting Reactor

Organic material, typically fed into the reactor at the top through a distribution mechanism, moves by gravity to an unloading mechanism at the bottom. Process control is usually by pressure-induced aeration, where the airflow is opposite to the downward materials flow. The height of these reactors makes process control difficult due to the high rates of airflow required per unit of distribution surface area. Neither temperature nor oxygen can be maintained at optimal levels throughout the reactors, leading to zones of non-optimal activity. As with static pile composting, a stable porous structure is important in vertical reactors, which usually lack internal mixing. Tall vertical reactors have been successfully used in the sludge composting industry where uniform feedstock and porous amendments can minimise these difficulties in process control, but are rarely used for heterogeneous materials like MSW.

- Horizontal composting reactors: These avoid high temperature, oxygen and moisture gradients of vertical reactors by maintaining a short airflow pathway. They come in a wide range of configurations, including static and agitated, pressure and/or vacuum-induced aeration. Agitated systems generally use the turning process to move the material through the system in a continuous mode, while static systems require a loading and unloading mechanism. Material handling equipment may also shred to a certain degree, exposing new surfaces for decomposition, but excessive shredding may also reduce porosity. Aeration systems are usually set in the floor of the reactor and may use temperature and/or oxygen as control variables. Figure illustrates a horizontal composting reactor:

Horizontal Composting Reactor

Systems with agitation and bed depths less than two to three meters appear effective in dealing with the heterogeneity of MSW.

Rotating drum: Rotating drum reactors take the trade-off between reactor cost and compost residence time to an even further extreme than the horizontal or vertical in-vessel systems. These reactors, also known as digesters, retain the material for only a few hours or days:

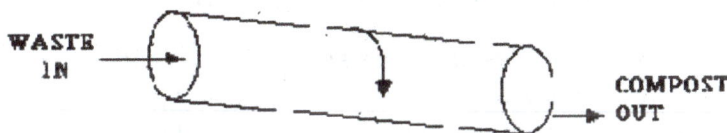

Rotating drum

While the tumbling action can help homogenise andshred materials,the short residence time usually means the processing is more physical than biological. While rotating drums can play an important role in MSW composting, they are normally followed by other biological processing, which may include in-vessel, static pile and/or windrow systems.

Anaerobic Composting

In anaerobic processes, facultative bacteria break down the organic materials in the absence of oxygen and produce methane and carbon dioxide. Anaerobic systems, if configured efficiently, will generate sufficient energy in the form of methane to operate the process and have enough surpluses to either market as gas or convert to electricity. Conventional composting systems, ontheother hand, need significant electrical or mechanical energy inputs to aerate or turn the piles. Several approaches are available for anaerobic digestion of feedstock.

Single-stage digesters contain the entire process in one airtight container. In this system, the feedstock is first shredded, and before being placed in the container,water and possibly nutrients are added to the previous ly shredded material. A single-stage digester may contain agitation equipment, which continuously stirs the liquefied material. The amount of water added and the presence or absence of agitation equipment depend on the particular research demonstration or proprietary process employed.

The two-stage digestion involves circulating a liquidsupernatantfromafirst stage digester, containing the materials, to a second-stage digester. This circulation eliminates the need for agitation equipment and provides the system operator with more opportunity to carefully control the biological process. Figure schematically illustrates an anaerobic digester with an aerobic compost curing:

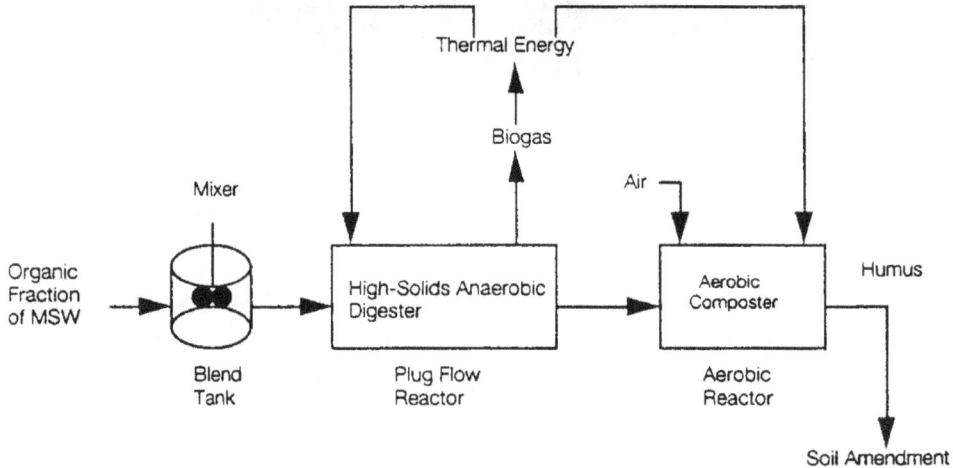

Anaerobic Digester with Aerobic Compost Curing

As digestion progresses, a mixture of methane and carbon dioxide is produced. These gases are continuously removed from both first- and second-stage digesters and are either combusted on-site or directed to off-site gas consumers. A portion of the recovered gas may be converted to thermal energy by combustion, which is then used to heat the digester. A stabilised residue remains, when the digestion process is completed. The residue is either removed from the digester with the mechanical equipment or pumped out as a liquid. It is chemically similar to compost, but contains more moisture. Conventional dewatering equipment can reduce the moisture content enough to handle the residue as a solid. The digested residue may require further curing by windrow or static pile composting.

Backyard Composting using dry leaves can be done with a little time and effort even with wastes from the kitchen. Any pile of waste organic matter will rot. To produce good compost for one's own garden following steps are to be followed.

• Choose a compost bin which can be of wood, cement, brick or plastic.

• Place the bin on a leveled, well-drained surface asit does not require base. Earthworm and other micro-organisms enter the bin from the base of bin through the soil and act on compost.

Note: Avoid wood bins that may be chemically treated to keep micro-organisms at bay.

• Carbon nitrogen ratio in the bin should be 25:1.

Note: Carbon-rich organic matters are brown colored waste products that include hay,sawdust,old-leaves,cardboard,newspaperandstraw.Nitrogen-richorganic matters are green colored products that include green grass, manure, coffee, weeds, and vegetable peels.

• Bottom layer of bin should be around six inches of carbon-rich matter like hay, straw etc.

Then top of it be covered with a three-inch layer of nitrogen-rich organic waste like vegetable scraps, manure etc.

- Water each layer in the bin.

- Add these layers alternately till it becomes say four to five feet high.

- Shred all the waste into smaller particles that you dump in the pile which makes decomposition process faster. A sample pile composition could be four parts of fruit and vegetable scraps, two parts of cow dung, one part of shredded paper, and one part dry leaves.

- Protect the pile from rain. To avoid sogginess of the contents cover the pile and see to it that contents are not too dry either.

- Enssure only biodegradable organic component are added.

- Proper decomposition occurs in presence of proper air and water. Don't saturate the pile with excess water. Turn the contents of the compost bin thoroughly using a shovel. This will aerate its contents, prevent it from stinking and hasten decomposition. It is advisable to mix the compost once in one or two weeks. Under proper temperature and moisture conditions, earthworms and other microbes will do the decomposition.

- If these steps are properly followed then would be ready in six months.

- Spread around two inches of compost on the soil until it mixeswell. Though the compost is spread on the top layer of the soil, it releases nutrients to the layers lying underneath.

Check your compost quality: Smell sweet like earth, dark in colour and crumbly.

Caution: Do not dump diseased or infected plants and vegetables as they may

survive even in the bin. Meat, dairy products, bones and fish attract rodents and other undesired creatures to your bin. Also, dog, cat or pig feces should not be put in the pile. However, chicken or cow dung is beneficial in a compost bin. Refrain from adding chemically-treated plants from your garden.

Monitoring compost: Touch the pile. If it is warm, it means microbes are doing

their job. If it is not warmer than temperature outside then feed kitchen waste and manure to the pile. It will increase decomposition. Keep the pile moist and not dry. Air is necessary for organisms working on your pile. Insert some tree branches in the pile so that they can be occasionally shaken to allow air circulation. Air and moisture act as pile's lifelines. If pile is still not decomposing, it means too much carbon and too little nitrogen is present in the pile.

The Indian Scenario

The first significant development in composting as a systemised process took place in India in 1925, when a process involving the anaerobic degradation of leaves, refuse, animal manure and sewerage were placed in pits. The materials were placed in layers and the pit wall conserved some of the heat of degradation, resulting in high temperature than when composting was carried out in the open. This process (often referred as Indore process) took approximately six months to produce usable compost. Following this, the Indian Council of Agricultural Research (ICAR) improved the method by laying down alternate layers of waste and sewage and this system (referred to as the Bangalore process) is still being used in India. In India, the high humid degradation that occursintheland requires a large amount of humus for maintaining soil-fertility, andforthat reason, composting is an ideal method for recycling organic wastes.

Biogasification

Biogas is a mixture of gases composed of methane (CH_4) 40 – 70 vol.%, carbon dioxide (CO_2) 30 – 60 vol.%, other gases 1 – 5 vol.% including, hydrogen (H_2) 0 – 1 vol.% and hydrogen sulphide (H_2S) 0 – 3 vol.%. It originates from bacteria in the process of bio-degradation of organic material under anaerobic (without air) conditions. The natural generation of biogasis an important part of the biogeochemical carbon cycle.

Methanogens (methane producing bacteria) are the last link in a chain of microorganisms, which degrade organic material and return the decomposition products to the environment. In thisprocess, biogasisgenerated, which isa source of renewable energy. As is the case with any pure gas, the characteristic properties of biogas are pressure and temperature dependency. It is also affected by moisture content. Well-functioning biogas systems can yield a whole range of benefitsfor their users, the society and the environment in general. Some of the important benefits are as follows:

- production of energy (heat, light, electricity);

- transformation of organic waste into high quality fertiliser;

- improvement of hygienic conditions through reduction of pathogens, worm eggs and flies;

- reduction of workload in firewood collection and cooking;

- environmental advantages through protection of soil, water, air and woody vegetation;

- micro-economical benefits through energy and fertilisersubstitution, additional income generation and increasing yields of animal husbandry and agriculture;

- macro-economic benefits through decentralised energy generation, import substitution and environmental protection.

What we can deduce from the list above, is that biogas technology can substantially contribute to energy conservation and development, if the economic viability and social acceptance of biogas technology are favourable.

Bio gasification or Bio methanation is the process of conversion of organic matter in the waste (liquid or solid) to Bio Methane (sometimes referred to as "Biogas" with high energy density) and manure (bio compost) by microbial action in the absence of air, known as "anaerobic processing or digestion."

Anaerobic Processing

Anaerobic processing of organic material is a two-stage process, where large organic polymers are fermented into short-chain volatile fatty acids. These acids are then converted into methane and carbon dioxide.

Biogasification of Organic Waste: Metabolic Stages

Note that both the organic polymers fermentation process and acid conversion occur at the same time, in a single-phase system. And, the separation of the acid-producing (acidogenic) bacteria from the-methaneproducing (methanogenic) bacteria results in a two-phase system (McDougall, et. al., 2001).

The main feature of anaerobic treatment is the concurrent waste stabilisation and production of methane gas, which is an energy source. The retention time for solid material in an anaerobic process can range from a few daysto several weeks, depending upon the chemical characteristics of solid material and the design of the biogasification system (e.g., single stage, two stage, multi-stage, wet or dry, temperature and pH control).

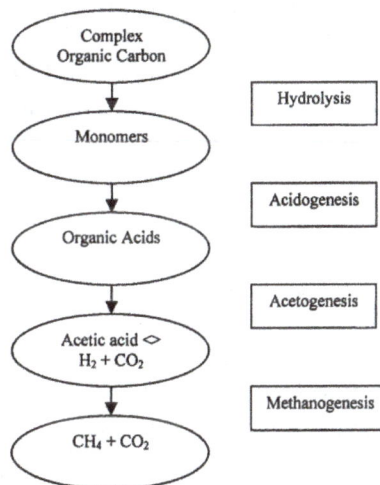

```
        Complex
     Organic Carbon
           │
           ▼                    Hydrolysis
        Monomers
           │
           ▼                    Acidogenesis
      Organic Acids
           │
           ▼                    Acetogenesis
       Acetic acid <>
         H₂ + CO₂
           │
           ▼                    Methanogenesis
        CH₄ + CO₂
```

In the absence of oxygen, anaerobic bacteria decompose organic matter as follows:

Organic matter + anaerobic bacteria $\rightarrow CH_4 + CO_2 + H_2S + NH_3$ + other end products + energy

The conditions for biogasification need to be anaerobic, for which a totally enclosed process vessel is required. Although this necessitates a higher level of technology than compared to composting, it allows a greater control over the process itself and the emission of noxious odours. Greater process control, especially of temperature, allows a reduction in treatment time, when compared to composting. Since a biogas plant is usually vertical, it also requires less area than a composting plant.

Biogasification is particularly suitable for wet substrates, such as sludges or food waste, which present difficulties in composting, as the lack of structural material restricts air circulation. The anaerobic process is used sometimes to digest sewage sludge, and this has been extended to frac-tionsofhouseholdsolid waste.

In contrast to aerobic processes (i.e., composting), biogasification is mildly exothermic. Thus, the heat needs to be supplied to maintain the process temperature, especially for thermophilic processes. The advantage of high temperature is that the reaction will occur at a faster rate, and so a shorter residence time is required in the reactor vessel.

According to the solid content of the material digested and the temperature at which the process operates, the various biogasification processes can be classified as under:

- Dry anaerobic digestion: In dry anaerobic digestion, semi-solid wastes are digested to produce biogas in a single stage, either as a batch process or a continuous process. It takes place at a total solid concentration of over 25%, and below this level of solid, the process is described as wet digestion. With regard to temperature, the processes are either described as mesophilic (operating between 30 and 40 C) or thermophilic (operating between 50 and 65 C), and anaerobic microorganisms have optimum growth rates within these temperature ranges. The dry fermentation process means that only little process water has to be added or heated, which favours thermophilic operation. No mixing equipment is necessary, and crust formation is not possible due to the relatively solid nature of the digester contents. This anaerobic process usually takes between 12 and 18 days, followed by several days in post-digestion for residue stabilisation and maturation.

- Wet anaerobic digestion: In its simplest form, this process consists of a single stage in a completely mixed mesophilic digester, operating at a total solid content of around 3 – 8%. To produce this level of dilution, a considerable amount of water has to be added and heated, and then removed after the digestion process. This method is routinely used to digest sewage sludge, and animal and household wastes. The single-stage wet process can suffer from several practical problems such as the formation of a hard scum layer in the digester and difficulty in keeping the content completely mixed.

The major problem with the single-stage process is that the different reactions in the process cannot be separately optimised. The acidogenic microorganisms grow fast and lower the pH of the reaction mixture, whereas the methanogens, which grow slowly, have a pH optimum around 7.0. The development of the two-stage digestion process solves this problem as hydrolysis and acidification occur in the first reactor vessel, kept at a pH of around 6.0 and methanogenesis occurs in the second vessel, operated at a pH of 7.5 – 8.2. The whole process can run with a retention time of 5 to 8 days.

Maturing or Refining

The residues of both wet and dry biogasification processes require extensive maturing under aerobic conditions. However, we can considerably reduce this period through effective aeration. The maturation processes facilitate the release of entrapped methane, elimination of phytotoxins (i.e., substances that are harmful for plant growth, such as volatile organic acids) and reduce the moisture content to an acceptable level. These residues contain a high level of water – even the dry process residue contains around 65% water.

Filtering or pressing can reduce excess water, and further drying can be achieved using waste heat from the gas engine, if the biogas isburnt onsite to produce electricity. The digested residue, initially anaerobic,willalsocontain many volatile organic acids and reduced organic materials. These need to be matured aerobically to oxidise and stabilise the compounds, intheprocess similar to the maturation of aerobic compost, prior to sale as compost or disposal as residue. Odour production is measured as the total amount of volatile organics produced per tonne of bio-waste during composting and the final aerobic maturation after anaerobic digestion.

Factors Affecting Biogasification

As with composting, a number of environmental factors (Phelps, et al., 1995) influence biogasification, some of which are listed below:

- Temperature: A temperature range of about 25 – 40 C (mesophilic) is generally optimal. It can be achieved without additional heating, thus being very economical. In some cases, additional energy input is provided to increase temperature to 50 – 60 C (thermophilic range) for greater gas production. Often digesters are constructed below ground to conserve heat.

- pH and alkalinity: pH close to neutral, i.e., 7, is optimum. At lower pH values (below 5.5), some bacteria carrying out the process areinhibited. Excess loading and presence of toxic materials will lower pH levels to below6.5 and can cause difficulties. When pH levels are too low, stopping the loading of the digester and/or use of time is recommended. The presence of alkalinity between 2500 and 5000 mg/L will provide good buffering against excessive pH changes.

- Nutrient concentration: An ideal C: N ratio of 25:1 is to be maintained in any digester. It is an important parameter, as anaerobic bacteria need nitrogen compound to grow and multiply. Too much nitrogen, however, can inhibit methanogenic activity. If the C: N ratio is high, then gas production can be enhanced by adding nitrogen, and if the C: N ratio is low, it can be increased by adding carbon, i.e., adding chicken manure, etc., which reduces the possibility of toxicity. Anaerobic digestion not only breaks down plant materials into biogas, but also releases plant nutrients, such as nitrogen (N), potassium (K) and phosphorous (P), and converts them into a form that can be easily absorbed by plants.

- Loading: When any digester is designed, the main variable to be defined is the internal volume. The digester volume is related to, as Fulford(1998) shows, two other parameters, and these are feed rate (Q, measuredin m³/day) and hydraulic loading or retention time (R, measured in days). The feed rate (Q) is given by the mass of total solid (m, kg) fed daily, divided by the proportion of total solid (TS) in the mixed slurry (as summing the density of feed is 1000 kg/m³).

$$Q \quad \frac{m}{TS \times 1000} = \frac{m}{TS\% \times 10} \quad m^3/day$$

The retention time (R) of any digester is given by the volume of the digester pit (V, m³), divided by the volume of the daily feed (Q, m³/day).

$$R = \frac{V}{Q} day$$

The loading rate, r (kg. VS/m³/day) of a digester is defined as the mass of volatile solids added each day per unit volume of digester. This is related to mass feed rate:

$$r = \frac{m \times VS}{Q} \quad or \quad r = \frac{m \times VS}{Q \times 100} \, kgVS / m \, / day$$

The typical values for the loading rate are between 0.2 and 2.0 kg VS/m 3/day.

Effect of toxins: The main cause of biogas plants receiving flak is the presence of toxic substance. Chlorinated hydrocarbons, such as chloroforms and other organic solvents, are particularly toxic to biogas digestion. If the digester hasbeen badly poisoned, it may be difficult to remove the toxins without removing most of the bacteria.In that case, the digester must be emptied, cleaned with plenty of water and refilled with fresh slurry.

Types of Digesters

In the anaerobic digestion process, the organic matter in mixtures of sludges is converted biologically, under anaerobic conditions, to a variety of products including methane (CH_4) and carbon dioxide (CO_2). The process is carried out in an airtight reactor, where wastes in the form of sludges are introduced continuously or intermittently and retained in the reactor for varying periods. The microbiology of anaerobic digestion and the optimum environmental considerations for the microorganisms can be achieved by selecting the proper type of digester.

Against this background, we explain below the operation and physical facilities for anaerobic digestion in single-stage digester (standard rate and high rate) and two-stage digester (Tchobanoglous and Burton, 1996), which generally operate in a mesophilic range, i.e., between 30 and 38 C:

(i) Standard rate single-stage digester: In a single stage digester, the untreated waste sludge is directly added to the zone, where the sludge is actively digested and the gas is being released. The sludge is heated by means of an external heat exchanger.

As the gas rises to the surface, it lifts sludge particles and other minerals such as grease oil and fats, ultimately giving rise to the formation of scum layer. As a result of digestion, the sludges stratify by forming a supernatant layer above the digesting sludge and become more mineralised, i.e., the percentage of fixed solid increases. Due to stratification and lack of mixing, the standard rate process is used principally for small installations. Detention time for standard rate processes vary from 30 to 60 days.

Figure: illustrates a single-stage digester:

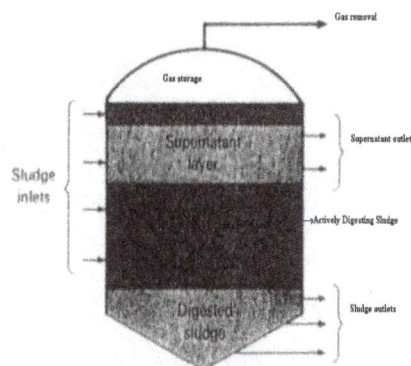

Standard Rate Single-Stage Digester

(ii) High rate single-stage digester: The single-stagehighratedigester differs from the single-stage standard rate digester in that the solid-loading rate is much greater. The sludge is mixed intimately by gas recirculation, mechanical mixing, pumping or draft tube mixer and heated to achieve optimum digestion rates.

With the exception of higher loading rates and improved mixing, there are only a few differences between standard rate and high rate digester. The mixing equipment should have a greater capacity and should reach the bottom of the tank, which should be deeper to aid the mixing process.

Digestion tank may have fixed roof or floating covers along with gasholder facility, which provide extra gas storage capacity. The required detention time for a high rate digestion is typically 15 days or less.

Figure: illustrates a high rate single-stage digester.

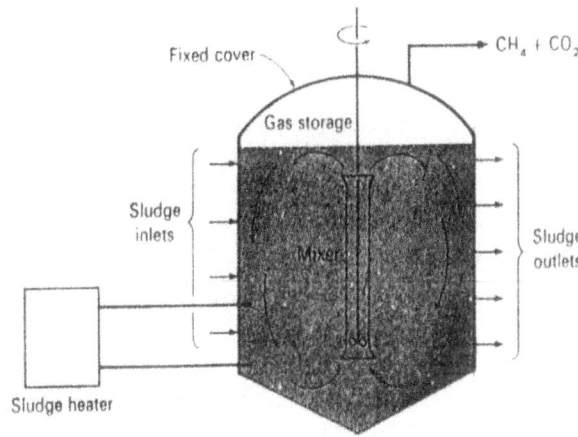

High Rate Single-Stage Digester

(iii) Two-stage digester: The combination of the two digesters, mentioned above, is known as a two-stage digester. The first stage digester is a high rate complete mix digester used for digestion, mixing and heating of waste sludge, while the primary function of a second stage isto separate the digested solid from the supernatant liquor, and in the process, additional digestion and gas production may occur. The tanks are made identical, in which case either one may be the primary digester. They may have fixed roofs or floating covers along with gasholder facility. Figure below illustrates a two-stage digester:

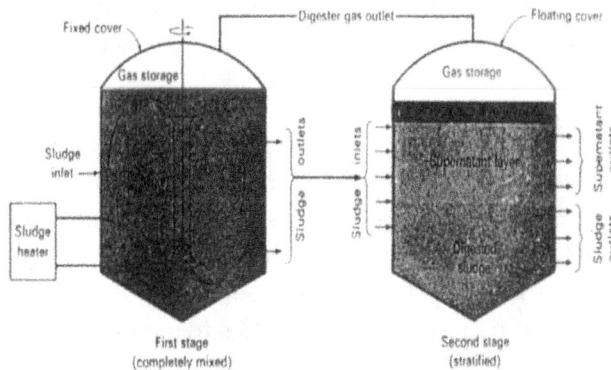

Two-Stage Digester

Mechanical Biological Treatment

A mechanical biological treatment (MBT) system is a type of waste processing facility that combines a sorting facility with a form of biological treatment such as composting or anaerobic digestion. MBT plants are designed to process mixed household waste as well as commercial and industrial wastes.

Process

Mixed waste input

Recyclables and/or ← Mechanical sorting → Rejects
refuse-derived fuel   & pre-treatment   to landfill

Compost/digestate ← Biological → Biogas & energy
or refuse derived   treatment   (anaerobic digestion)

Process flow chart

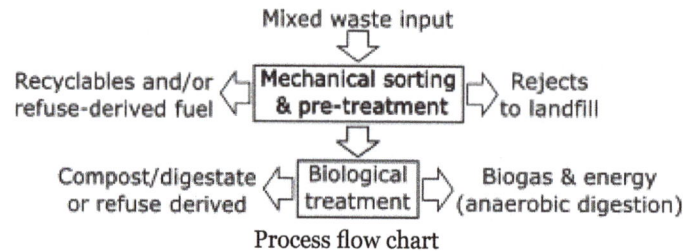

The terms *mechanical biological treatment* or *mechanical biological pre-treatment* relate to a group of solid waste treatment systems. These systems enable the recovery of materials contained within the mixed waste and facilitate the stabilisation of the biodegradable component of the material.

The sorting component of the plants typically resemble a materials recovery facility. This component is either configured to recover the individual elements of the waste or produce a Refuse-derived fuel that can be used for the generation of power.

The components of the mixed waste stream that can be recovered include:

- Ferrous Metal

- Non-ferrous metal

- Plastic

- Glass

Terminology

MBT is also sometimes termed BMT – biological mechanical treatment – however this simply refers to the order of processing, i.e. the biological phase of the system precedes the mechanical sorting. MBT should not be confused with MHT – *mechanical heat treatment*

Mechanical Sorting

The "mechanical" element is usually an automated mechanical sorting stage. This either removes recyclable elements from a mixed waste stream (such as metals, plastics, glass and paper) or processes them. It typically involves factory style conveyors, industrial magnets, eddy current separators, trommels, shredders and other tailor made systems, or the sorting is done manually at hand picking stations. The mechanical element has a number of similarities to a materials recovery facility (MRF).

Wet material recovery facility, Hiriya, Israel

Some systems integrate a wet MRF to separate by density and floatation and to recover & wash the recyclable elements of the waste in a form that can be sent for recycling. MBT can alternatively process the waste to produce a high calorific fuel termed refuse derived fuel (RDF). RDF can be used in cement kilns or thermal combustion power plants and is generally made up from plastics and biodegradable organic waste. Systems which are configured to produce RDF include the Herhof and Ecodeco Processes. It is a common misconception that all MBT processes produce RDF. This is not the case and depends strictly on system configuration and suitable local markets for MBT outputs.

Biological Processing

Twin stage & UAS Banaerobic digesters

The "biological" element refers to either:

- Anaerobic digestion

- Composting

- Biodrying

Anaerobic digestion harnesses anaerobic microorganisms to break down the biodegradable component of the waste to produce biogas and soil improver. The biogas can be used to generate electricity and heat.

Biological can also refer to a composting stage. Here the organic component is broken down by naturally occurring aerobic microorganisms. They breakdown the waste into carbon dioxide and compost. There is no green energy produced by systems employing only composting treatment for the biodegradable waste.

In the case of biodrying, the waste material undergoes a period of rapid heating through the action of aerobic microbes. During this partial composting stage the heat generated by the microbes result in rapid drying of the waste. These systems are often configured to produce a refuse-derived fuel where a dry, light material is advantageous for later transport and combustion.

Some systems incorporate both anaerobic digestion and composting. This may either take the form of a full anaerobic digestion phase, followed by the maturation (composting) of the digestate. Alternatively a partial anaerobic digestion phase can be induced on water that is percolated through the raw waste, dissolving the readily available sugars, with the remaining material being sent to a windrow composting facility.

By processing the biodegradable waste either by anaerobic digestion or by composting MBT technologies help to reduce the contribution of greenhouse gases to global warming.

Usable wastes for this system:

- Municipal solid waste

- Commercial and industrial waste

- Sewage sludge

Possible products of this system:

- Renewable fuel (biogas) leading to renewable power

- Recovered recyclable materials such as metals, paper, plastics, glass etc.

- Digestate - an organic fertiliser and soil improver

- Carbon credits – additional revenues

- High calorific fraction refuse derived fuel - Renewable fuel content dependent upon biological component

- Residual unusable materials prepared for their final safe treatment (e.g. incineration or gasification) and/or landfill

Further advantages:

- Small fraction of inert residual waste

- Reduction of the waste volume to be deposited to at least a half (density > 1.3 t/m^3), thus the lifetime of the landfill is at least twice as long as usually

- Utilisation of the leachate in the process

- Landfill gas not problematic as biological component of waste has been stabilised

- Daily covering of landfill not necessary

Consideration of Applications

MBT systems can form an integral part of a region's waste treatment infrastructure. These systems are typically integrated with kerbside collection schemes. In the event that a refuse-derived fuel is produced as a by-product then a combustion facility would be required. This could either be an incineration facility or a gasifier.

Alternatively MBT solutions can diminish the need for home separation and kerbside collection of recyclable elements of waste. This gives the ability of local authorities, municipalities and councils to reduce the use of waste vehicles on the roads and keep recycling rates high.

Position of Environmental Groups

Friends of the Earth suggests that the best environmental route for residual waste is to firstly maximise removal of remaining recyclable materials from the waste stream (such as metals, plastics and paper). The amount of waste remaining should be composted or anaerobically digested and disposed of to landfill, unless sufficiently clean to be used as compost.

A report by Eunomia undertook a detailed analysis of the climate impacts of different residual waste technologies. It found that an MBT process that extracts both the metals and plastics prior to landfilling is one of the best options for dealing with our residual waste, and has a lower impact than either MBT processes producing RDF for incineration or incineration of waste without MBT.

Friends of the Earth does not support MBT plants that produce refuse derived fuel (RDF), and believes MBT processes should occur in small, localised treatment plants.

Anaerobic Digestion

Anaerobic digestion is a collection of processes by which microorganisms break down biodegradable material in the absence of oxygen. The process is used for industrial or domestic purposes to manage waste or to produce fuels. Much of the fermentation used industrially to produce food and drink products, as well as home fermentation, uses anaerobic digestion.

Anaerobic digestion occurs naturally in some soils and in lake and oceanic basin sediments, where it is usually referred to as "anaerobic activity". This is the source of marsh gas methane as discovered by Volta in 1776.

The digestion process begins with bacterialhydrolysis of the input materials. Insoluble organic polymers, such as carbohydrates, are broken down to soluble derivatives that become available for other bacteria. Acidogenic bacteria then convert the sugars and amino acids into carbon dioxide, hydrogen, ammonia, and organic acids. These bacteria convert these resulting organic acids into acetic acid, along with additional ammonia, hydrogen, and carbon dioxide. Finally, methanogens convert these products to methane and carbon dioxide. The methanogenic archaea populations play an indispensable role in anaerobic wastewater treatments.

Anaerobic digestion is used as part of the process to treat biodegradable waste and sewage sludge. As part of an integrated waste management system, anaerobic digestion reduces the emission of landfill gas into the atmosphere. Anaerobic digesters can also be fed with purpose-grown energy crops, such as maize.

Anaerobic digestion is widely used as a source of renewable energy. The process produces a biogas, consisting of methane, carbon dioxide and traces of other 'contaminant' gases. This biogas can be used directly as fuel, in combined heat and power gas engines or upgraded to natural gas-quality biomethane. The nutrient-rich digestate also produced can be used as fertilizer.

With the re-use of waste as a resource and new technological approaches that have lowered capital costs, anaerobic digestion has in recent years received increased attention among governments in a number of countries, among these the United Kingdom (2011), Germany and Denmark (2011).

Process

Many microorganisms affect anaerobic digestion, including acetic acid-forming bacteria (acetogens) and methane-forming archaea (methanogens). These organisms promote a number of chemical processes in converting the biomass to biogas.

Gaseous oxygen is excluded from the reactions by physical containment. Anaerobes utilize electron acceptors from sources other than oxygen gas. These acceptors can be the organic material itself or may be supplied by inorganic oxides from within the input material. When the oxygen source in an anaerobic system is derived from the organic material itself, the 'intermediate' end products are primarily alcohols, aldehydes, and organic acids, plus carbon dioxide. In the presence of specialised methanogens, the intermediates are converted to the 'final' end products of methane, carbon dioxide, and trace levels of hydrogen sulfide. In an anaerobic system, the majority of the chemical energy contained within the starting material is released by methanogenic bacteria as methane.

Populations of anaerobic microorganisms typically take a significant period of time to establish themselves to be fully effective. Therefore, common practice is to introduce anaerobic microorganisms from materials with existing populations, a process known as "seeding" the digesters, typically accomplished with the addition of sewage sludge or cattle slurry.

Process Stages

The four key stages of anaerobic digestion involve hydrolysis, acidogenesis, acetogenesis and methanogenesis. The overall process can be described by the chemical reaction, where organic material such as glucose is biochemically digested into carbon dioxide (CO_2) and methane (CH_4) by the anaerobic microorganisms.

$$C_6H_{12}O_6 \rightarrow 3CO_2 + 3CH_4$$

- Hydrolysis

In most cases, biomass is made up of large organic polymers. For the bacteria in anaerobic digesters to access the energy potential of the material, these chains must first be broken down into their smaller constituent parts. These constituent parts, or monomers, such as sugars, are readily available to other bacteria. The process of breaking these chains and dissolving the smaller molecules into solution is called hydrolysis. Therefore, hydrolysis of these high-molecular-weight polymeric components is the necessary first step in anaerobic digestion. Through hydrolysis the complex organic molecules are broken down into simple sugars, amino acids, and fatty acids.

Acetate and hydrogen produced in the first stages can be used directly by methanogens. Other molecules, such as volatile fatty acids (VFAs) with a chain length greater than that of acetate must first be catabolised into compounds that can be directly used by methanogens.

- Acidogenesis

The biological process of acidogenesis results in further breakdown of the remaining components by acidogenic (fermentative) bacteria. Here, VFAs are created, along with ammonia, carbon dioxide, and hydrogen sulfide, as well as other byproducts. The process of acidogenesis is similar to the way milk sours.

- Acetogenesis

The third stage of anaerobic digestion is acetogenesis. Here, simple molecules created through the acidogenesis phase are further digested by acetogens to produce largely acetic acid, as well as carbon dioxide and hydrogen.

- Methanogenesis

The terminal stage of anaerobic digestion is the biological process of methanogenesis. Here, methanogens use the intermediate products of the preceding stages and convert them into methane, carbon dioxide, and water. These components make up the majority of the biogas emitted from the system. Methanogenesis is sensitive to both high and low pHs and occurs between pH 6.5 and pH 8. The remaining, indigestible material the microbes cannot use and any dead bacterial remains constitute the digestate.

Configuration

Anaerobic digesters can be designed and engineered to operate using a number of different configurations and can be categorized into batch vs. continuous process mode, mesophilic vs. thermophilic temperature conditions, high vs. low portion of solids, and single stage vs. multistage processes. More initial build money and a larger volume of the batch digester is needed to handle the same amount of waste as a continuous process digester. Higher heat energy is demanded in a thermophilic system compared to a mesophilic system and has a larger gas output capacity and higher methane gas content. For solids content, low will handle up to 15% solid content. Above this level is considered high solids content and can also be known as dry digestion. In a single stage process, one reactor houses the four anaerobic digestion steps. A multistage process utilizes two or more reactors for digestion to separate the methanogenesis and hydrolysis phases.

Batch or continuous

Anaerobic digestion can be performed as a batch process or a continuous process. In a batch system, biomass is added to the reactor at the start of the process. The reactor is then sealed for the duration of the process. In its simplest form batch processing needs inoculation with already processed material to start the anaerobic digestion. In a typical scenario, biogas production will be formed with a normal distribution pattern over time. Operators can use this fact to determine when they believe the process of digestion of the organic matter has completed. There can be severe odour issues if a batch reactor is opened and emptied before the process is well completed. A more advanced type of batch approach has limited the odour issues by integrating anaerobic digestion with in-vessel composting. In this approach inoculation takes place through the use of recirculated degasified percolate. After anaerobic digestion has completed, the biomass is kept in the reactor which is then used for in-vessel composting before it is opened As the batch digestion is simple and requires less equipment and lower levels of design work, it is typically a cheaper form of digestion. Using more than one batch reactor at a plant can ensure constant production of biogas.

In continuous digestion processes, organic matter is constantly added (continuous complete mixed) or added in stages to the reactor (continuous plug flow; first in – first out). Here, the end products are constantly or periodically removed, resulting in constant production of biogas. A single or multiple digesters in sequence may be used. Examples of this form of anaerobic digestion include continuous stirred-tank reactors, upflow anaerobic sludge blankets, expanded granular sludge beds and internal circulation reactors.

Temperature

The two conventional operational temperature levels for anaerobic digesters determine the species of methanogens in the digesters:

- *Mesophilic* digestion takes place optimally around 30 to 38 °C, or at ambient temperatures between 20 and 45 °C, where mesophiles are the primary microorganism present.

- *Thermophilic* digestion takes place optimally around 49 to 57 °C, or at elevated temperatures up to 70 °C, where thermophiles are the primary microorganisms present.

A limit case has been reached in Bolivia, with anaerobic digestion in temperature working conditions of less than 10 °C. The anaerobic process is very slow, taking more than three times the normal mesophilic time process. In experimental work at University of Alaska Fairbanks, a 1,000 litre digester using psychrophiles harvested from "mud from a frozen lake in Alaska" has produced 200–300 litres of methane per day, about 20 to 30% of the output from digesters in warmer climates. Mesophilic species outnumber thermophiles, and they are also more tolerant to changes in environmental conditions than thermophiles. Mesophilic systems are, therefore, considered to be more stable than thermophilic digestion systems. In contrast, while thermophilic digestion systems are considered less stable, their energy input is higher, with more biogas being removed from the organic matter in an equal amount of time. The increased temperatures facilitate faster reaction rates, and thus faster gas yields. Operation at higher temperatures facilitates greater pathogen reduction of the digestate. In countries where legislation, such as the Animal By-Products Regulations in the European Union, requires digestate to meet certain levels of pathogen reduction there may be a benefit to using thermophilic temperatures instead of mesophilic.

Additional pre-treatment can be used to reduce the necessary retention time to produce biogas. For example, certain processes shred the substrates to increase the surface area or use a thermal pretreatment stage (such as pasteurisation) to significantly enhance the biogas output. The pasteurisation process can also be used to reduce the pathogenic concentration in the digesate leaving the anaerobic digester. Pasteurisation may be achieved by heat treatment combined with maceration of the solids.

Solids Content

In a typical scenario, three different operational parameters are associated with the solids content of the feedstock to the digesters:

- High solids (dry—stackable substrate)

- High solids (wet—pumpable substrate)

- Low solids (wet—pumpable substrate)

High solids (dry) digesters are designed to process materials with a solids content between 25 and 40%. Unlike wet digesters that process pumpable slurries, high solids (dry – stackable substrate) digesters are designed to process solid substrates without the addition of water. The primary styles of dry digesters are continuous vertical plug flow and batch tunnel horizontal digesters. Continuous vertical plug flow digesters are upright, cylindrical tanks where feedstock is continuously fed into the top of the digester, and flows downward by gravity during digestion. In batch tunnel digesters, the feedstock is deposited in tunnel-like chambers with a gas-tight door. Neither approach has mixing inside the digester. The amount of pretreatment, such as contaminant removal, depends both upon the nature of the waste streams being processed and the desired quality of the digestate. Size reduction (grinding) is beneficial in continuous vertical systems, as it accelerates digestion, while batch systems avoid grinding and instead require structure (e.g. yard waste) to reduce compaction of the stacked pile. Continuous vertical dry digesters have a smaller footprint due to the shorter effective retention time and vertical design. Wet digesters can be designed to operate in either a high-solids content, with a total suspended solids (TSS) concentration greater than ~20%, or a low-solids concentration less than ~15%.

High solids (wet) digesters process a thick slurry that requires more energy input to move and process the feedstock. The thickness of the material may also lead to associated problems with abrasion. High solids digesters will typically have a lower land requirement due to the lower volumes associated with the moisture. High solids digesters also require correction of conventional performance calculations (e.g. gas production, retention time, kinetics, etc.) originally based on very dilute sewage digestion concepts, since larger fractions of the feedstock mass are potentially convertible to biogas.

Low solids (wet) digesters can transport material through the system using standard pumps that require significantly lower energy input. Low solids digesters require a larger amount of land than high solids due to the increased volumes associated with the increased liquid-to-feedstock ratio of the digesters. There are benefits associated with operation in a liquid environment, as it enables more thorough circulation of materials and contact between the bacteria and their food. This enables the bacteria to more readily access the substances on which they are feeding, and increases the rate of gas production.

Complexity

Digestion systems can be configured with different levels of complexity. In a single-stage digestion system (one-stage), all of the biological reactions occur within a single, sealed reactor or holding tank. Using a single stage reduces construction costs, but results in less control of the reactions occurring within the system. Acidogenic bacteria, through the production of acids, reduce the pH of the tank. Methanogenic bacteria, as outlined earlier, operate in a strictly defined pH range. Therefore, the biological reactions of the different species in a single-stage reactor can be in direct competition with each other. Another one-stage reaction system is an anaerobic lagoon. These lagoons are pond-like, earthen basins used for the treatment and long-term storage of manures. Here the anaerobic reactions are contained within the natural anaerobic sludge contained in the pool.

In a two-stage digestion system (multistage), different digestion vessels are optimised to bring maximum control over the bacterial communities living within the digesters. Acidogenic bacteria produce organic acids and more quickly grow and reproduce than methanogenic bacteria. Methanogenic bacteria require stable pH and temperature to optimise their performance.

Under typical circumstances, hydrolysis, acetogenesis, and acidogenesis occur within the first reaction vessel. The organic material is then heated to the required operational temperature (either mesophilic or thermophilic) prior to being pumped into a methanogenic reactor. The initial hydrolysis or acidogenesis tanks prior to the methanogenic reactor can provide a buffer to the rate at which feedstock is added. Some European countries require a degree of elevated heat treatment to kill harmful bacteria in the input waste. In this instance, there may be a pasteurisation or sterilisation stage prior to digestion or between the two digestion tanks. Notably, it is not possible to completely isolate the different reaction phases, and often some biogas is produced in the hydrolysis or acidogenesis tanks.

Residence Time

The residence time in a digester varies with the amount and type of feed material, and with the configuration of the digestion system. In a typical two-stage mesophilic digestion, residence time varies between 15 and 40 days, while for a single-stage thermophilic digestion, residence times is normally faster and takes around 14 days. The plug-flow nature of some of these systems will mean the full degradation of the material may not have been realised in this timescale. In this event, digestate exiting the system will be darker in colour and will typically have more odour.

In the case of an upflow anaerobic sludge blanket digestion (UASB), hydraulic residence times can be as short as 1 hour to 1 day, and solid retention times can be up to 90 days. In this manner, a UASB system is able to separate solids and hydraulic retention times with the use of a sludge blanket. Continuous digesters have mechanical or hydraulic devices, depending on the level of solids in the material, to mix the contents, enabling the bacteria and the food to be in contact. They also allow excess material to be continuously extracted to maintain a reasonably constant volume within the digestion tanks.

Inhibition

The anaerobic digestion process can be inhibited by several compounds, affecting one or more of the bacterial groups responsible for the different organic matter degradation steps. The degree of the inhibition depends, among other factors, on the concentration of the inhibitor in the digester. Potential inhibitors are ammonia, sulfide, light metal ions (Na, K, Mg, Ca, Al), heavy metals, some organics (chlorophenols, halogenated aliphatics, N-substituted aromatics, long chain fatty acids), etc.

Top: Farm-based maize silage digester located near Neumünster in Germany, 2007 - the green, inflatable biogas holder is shown on top of the digester. *Bottom:* Two-stage, low solids, UASB digestion component of a mechanical biological treatment system near Tel Aviv; the process water is seen in balance tank and sequencing batch reactor, 2005.

Feedstocks

Anaerobic lagoon and generators at the Cal Poly Dairy, United States

The most important initial issue when considering the application of anaerobic digestion systems is the feedstock to the process. Almost any organic material can be processed with anaerobic digestion; however, if biogas production is the aim, the level of putrescibility is the key factor in its successful application. The more putrescible (digestible) the material, the higher the gas yields possible from the system.

Feedstocks can include biodegradable waste materials, such as waste paper, grass clippings, leftover food, sewage, and animal waste. Woody wastes are the exception, because they are largely unaffected by digestion, as most anaerobes are unable to degrade lignin. Xylophalgeous anaerobes (lignin consumers) or using high temperature pretreatment, such as pyrolysis, can be used to break down the lignin. Anaerobic digesters can also be fed with specially grown energy crops, such as silage, for dedicated biogas production. In Germany and continental Europe, these facilities are referred to as "biogas" plants. A codigestion or cofermentation plant is typically an agricultural anaerobic digester that accepts two or more input materials for simultaneous digestion.

The length of time required for anaerobic digestion depends on the chemical complexity of the material. Material rich in easily digestible sugars breaks down quickly where as intact lignocellulosic material rich in cellulose and hemicellulose polymers can take much longer to break down. Anaerobic microorganisms are generally unable to break down lignin, the recalcitrant aromatic component of biomass.

Anaerobic digesters were originally designed for operation using sewage sludge and manures. Sewage and manure are not, however, the material with the most potential for anaerobic digestion, as the biodegradable material has already had much of the energy content taken out by the animals that produced it. Therefore, many digesters operate with codigestion of two or more types of feedstock. For example, in a farm-based digester that uses dairy manure as the primary feedstock, the gas production may be significantly increased by adding a second feedstock, e.g., grass and corn (typical on-farm feedstock), or various organic byproducts, such as slaughterhouse waste, fats, oils and grease from restaurants, organic household waste, etc. (typical off-site feedstock).

Digesters processing dedicated energy crops can achieve high levels of degradation and biogas production. Slurry-only systems are generally cheaper, but generate far less energy than those using crops, such as maize and grass silage; by using a modest amount of crop material (30%), an anaerobic digestion plant can increase energy output tenfold for only three times the capital cost, relative to a slurry-only system.

Moisture Content

A second consideration related to the feedstock is moisture content. Dryer, stackable substrates, such as food and yard waste, are suitable for digestion in tunnel-like chambers. Tunnel-style systems typically have near-zero wastewater discharge, as well, so this style of system has advantages where the discharge of digester liquids are a liability. The wetter the material, the more suitable it will be to handling with standard pumps instead of energy-intensive concrete pumps and physical means of movement. Also, the wetter the material, the more volume and area it takes up relative to the levels of gas produced. The moisture content of the target feedstock will also affect what type of system is applied to its treatment. To use a high-solids anaerobic digester for dilute feedstocks, bulking agents, such as compost, should be applied to increase the solids content of the input material. Another key consideration is the carbon:nitrogen ratio of the input material. This ratio is the balance of food a microbe requires to grow; the optimal C:N ratio is 20–30:1. Excess N can lead to ammonia inhibition of digestion.

Contamination

The level of contamination of the feedstock material is a key consideration. If the feedstock to the digesters has significant levels of physical contaminants, such as plastic, glass, or metals, then processing to remove the contaminants will be required for the material to be used. If it is not removed, then the digesters can be blocked and will not function efficiently. It is with this understanding that mechanical biological treatment plants are designed. The higher the level of pretreatment a feedstock requires, the more processing machinery will be required, and, hence, the project will have higher capital costs.

After sorting or screening to remove any physical contaminants from the feedstock, the material is often shredded, minced, and mechanically or hydraulically pulped to increase the surface area available to microbes in the digesters and, hence, increase the speed of digestion. The maceration of solids can be achieved by using a chopper pump to transfer the feedstock material into the airtight digester, where anaerobic treatment takes place.

Substrate Composition

Substrate composition is a major factor in determining the methane yield and methane production rates from the digestion of biomass. Techniques to determine the compositional characteristics of the feedstock are available, while parameters such as solids, elemental, and organic analyses are important for digester design and operation.

Applications

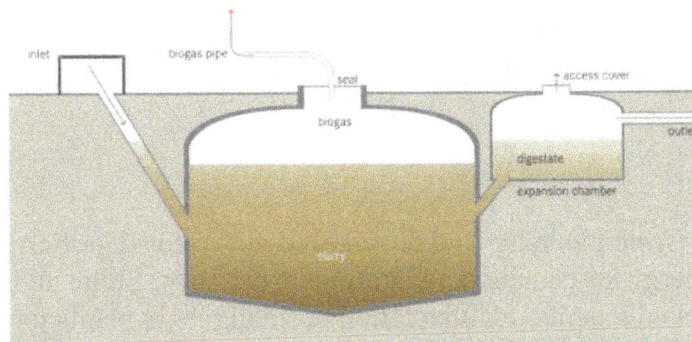

Schematic of an anaerobic digester as part of a sanitation system. It produces a digested slurry (digestate) that can be used as a fertilizer, and biogas that can be used for energy.

Using anaerobic digestion technologies can help to reduce the emission of greenhouse gases in a number of key ways:

- Replacement of fossil fuels

- Reducing or eliminating the energy footprint of waste treatment plants

- Reducing methane emission from landfills

- Displacing industrially produced chemical fertilizers

- Reducing vehicle movements

- Reducing electrical grid transportation losses

- Reducing usage of LP Gas for cooking

Waste and Wastewater Treatment

Anaerobic digestion is particularly suited to organic material, and is commonly used for industrial effluent, wastewater and sewage sludge treatment. Anaerobic digestion, a simple process, can greatly reduce the amount of organic matter which might otherwise be destined to be dumped at sea, dumped in landfills, or burnt in incinerators.

Pressure from environmentally related legislation on solid waste disposal methods in developed countries has increased the application of anaerobic digestion as a process for reducing waste volumes and generating useful byproducts. It may either be used to process the source-separated fraction of municipal waste or alternatively combined with mechanical sorting systems, to process residual mixed municipal waste. These facilities are called mechanical biological treatment plants.

If the putrescible waste processed in anaerobic digesters were disposed of in a landfill, it would break down naturally and often anaerobically. In this case, the gas will eventually escape into the atmosphere. As methane is about 20 times more potent as a greenhouse gas than carbon dioxide, this has significant negative environmental effects.

In countries that collect household waste, the use of local anaerobic digestion facilities can help to reduce the amount of waste that requires transportation to centralized landfill sites or incineration facilities. This reduced burden on transportation reduces carbon emissions from the collection vehicles. If localized anaerobic digestion facilities are embedded within an electrical distribution network, they can help reduce the electrical losses associated with transporting electricity over a national grid.

Power Generation

In developing countries, simple home and farm-based anaerobic digestion systems offer the potential for low-cost energy for cooking and lighting. From 1975, China and India have both had large, government-backed schemes for adaptation of small biogas plants for use in the household for cooking and lighting. At present, projects for anaerobic digestion in the developing world can gain financial support through the United NationsClean Development Mechanism if they are able to show they provide reduced carbon emissions.

Methane and power produced in anaerobic digestion facilities can be used to replace energy derived from fossil fuels, and hence reduce emissions of greenhouse gases, because the carbon in biodegradable material is part of a carbon cycle. The carbon released into the atmosphere from the combustion of biogas has been removed by plants for them to grow in the recent past, usually within the last decade, but more typically within the last growing season. If the plants are regrown, taking the carbon out of the atmosphere once more, the system will be carbon neutral. In contrast, carbon in fossil fuels has been sequestered in the earth for many millions of years, the combustion of which increases the overall levels of carbon dioxide in the atmosphere.

Biogas from sewage works is sometimes used to run a gas engine to produce electrical power, some or all of which can be used to run the sewage works. Some waste heat from the engine is then used to heat the digester. The waste heat is, in general, enough to heat the digester to the required temperatures. The power potential from sewage works is limited – in the UK, there are about 80 MW total of such generation, with the potential to increase to 150 MW, which is insignificant compared to the average power demand in the UK of about 35,000 MW. The scope for biogas generation from nonsewage waste biological matter – energy crops, food waste, abattoir waste, etc. - is much higher, estimated to be capable of about 3,000 MW. Farm biogas plants using animal waste and energy crops are expected to contribute to reducing CO_2 emissions and strengthen the grid, while providing UK farmers with additional revenues.

Some countries offer incentives in the form of, for example, feed-in tariffs for feeding electricity onto the power grid to subsidize green energy production.

In Oakland, California at the East Bay Municipal Utility District's main wastewater treatment plant (EBMUD), food waste is currently codigested with primary and secondary municipal wastewater solids and other high-strength wastes. Compared to municipal wastewater solids digestion alone, food waste codigestion has many benefits. Anaerobic digestion of food waste pulp from the EBMUD food waste process provides a higher normalized energy benefit, compared to municipal wastewater solids: 730 to 1,300 kWh per dry ton of food waste applied compared to 560 to 940 kWh per dry ton of municipal wastewater solids applied.

Grid Injection

Biogas grid-injection is the injection of biogas into the natural gas grid. The raw biogas has to be previously upgraded to biomethane. This upgrading implies the removal of contaminants such as hydrogen sulphide or siloxanes, as well as the carbon dioxide. Several technologies are available for this purpose, being the most widely implemented the pressure swing adsorption (PSA), water or amine scrubbing (absorption processes) and, in the last years, membrane separation. As an alternative, the electricity and the heat can be used for on-site generation, resulting in a reduction of losses in the transportation of energy. Typical energy losses in natural gas transmission systems range from 1–2%, whereas the current energy losses on a large electrical system range from 5–8%.

In October 2010, Didcot Sewage Works became the first in the UK to produce biomethane gas supplied to the national grid, for use in up to 200 homes in Oxfordshire. By 2017, UK electricity firm Ecotricity plan to have digester fed by locally sourced grass fueling 6000 homes

Vehicle Fuel

After upgrading with the above-mentioned technologies, the biogas (transformed into biomethane) can be used as vehicle fuel in adapted vehicles. This use is very extensive in Sweden, where over 38,600 gas vehicles exist, and 60% of the vehicle gas is biomethane generated in anaerobic digestion plants.

Fertiliser and Soil Conditioner

The solid, fibrous component of the digested material can be used as a soil conditioner to increase the organic content of soils. Digester liquor can be used as a fertiliser to supply vital nutrients to soils instead of chemical fertilisers that require large amounts of energy to produce and transport. The use of manufactured fertilisers is, therefore, more carbon-intensive than the use of anaerobic digester liquor fertiliser. In countries such as Spain, where many soils are organically depleted, the markets for the digested solids can be equally as important as the biogas.

Cooking Gas

By using a bio-digester, which produces the bacteria required for decomposing, cooking gas is generated. The organic garbage like fallen leaves, kitchen waste, food waste etc. are fed into a crusher unit, where the mixture is conflated with a small amount of water. The mixture is then fed into the bio-digester, where the bacteria decomposes it to produce cooking gas. This gas is piped to kitchen stove. A 2 cubic meter bio-digester can produce 2 cubic meter of cooking gas. This is equivalent to 1 kg of LPG. The notable advantage of using a bio-digester is the sludge which is a rich organic manure.

Products

The three principal products of anaerobic digestion are biogas, digestate, and water.

Biogas

Typical Composition of Biogas		
Compound	**Formula**	**%**
Methane	CH_4	50–75
Carbon dioxide	CO_2	25–50
Nitrogen	N_2	0–10
Hydrogen	H_2	0–1
Hydrogen sulphide	H_2S	0–3
Oxygen	O_2	0–0

Biogas is the ultimate waste product of the bacteria feeding off the input biodegradable feedstock (the methanogenesis stage of anaerobic digestion is performed by archaea (a micro-organism on a distinctly different branch of the phylogenetic tree of life to bacteria), and is mostly methane and carbon dioxide, with a small amount hydrogen and trace hydrogen sulfide. (As-produced, biogas also contains water vapor, with the fractional water vapor volume a function of biogas temperature). Most of the biogas is produced during the middle of the digestion, after the bacterial population has grown, and tapers off as the putrescible material is exhausted. The gas is normally stored on top of the digester in an inflatable gas bubble or extracted and stored next to the facility in a gas holder.

The methane in biogas can be burned to produce both heat and electricity, usually with a reciprocating engine or microturbine often in a cogeneration arrangement where the electricity and waste heat generated are used to warm the digesters or to heat buildings. Excess electricity can be sold to suppliers or put into the local grid. Electricity produced by anaerobic digesters is considered to be renewable energy and may attract subsidies. Biogas does not contribute to increasing atmospheric carbon dioxide concentrations because the gas is not released directly into the atmosphere and the carbon dioxide comes from an organic source with a short carbon cycle.

Biogas may require treatment or 'scrubbing' to refine it for use as a fuel. Hydrogen sulfide, a toxic product formed from sulfates in the feedstock, is released as a trace component of the biogas. National environmental enforcement agencies, such as the U.S. Environmental Protection Agency or the English and Welsh Environment Agency, put strict limits on the levels of gases containing hydrogen sulfide, and, if the levels of hydrogen sulfide in the gas are high, gas scrubbing and cleaning equipment (such as amine gas treating) will be needed to process the biogas to within regionally accepted levels. Alternatively, the addition of ferrous chloride $FeCl_2$ to the digestion tanks inhibits hydrogen sulfide production.

Volatilesiloxanes can also contaminate the biogas; such compounds are frequently found in household waste and wastewater. In digestion facilities accepting these materials as a component of the feedstock, low-molecular-weight siloxanes volatilise into biogas. When this gas is combusted in a gas engine, turbine, or boiler, siloxanes are converted into silicon dioxide (SiO_2), which deposits internally in the machine, increasing wear and tear. Practical and cost-effective technologies to remove siloxanes and other biogas contaminants are available at the present time. In certain applications, *in situ* treatment can be used to increase the methane purity by reducing the offgas carbon dioxide content, purging the majority of it in a secondary reactor.

In countries such as Switzerland, Germany, and Sweden, the methane in the biogas may be compressed for it to be used as a vehicle transportation fuel or input directly into the gas mains. In countries where the driver for the use of anaerobic digestion are renewable electricity subsidies, this route of treatment is less likely, as energy is required in this processing stage and reduces the overall levels available to sell.

Biogas holder with lightning protection rods and backup gas flare

Biogas carrying pipes

Digestate

Digestate is the solid remnants of the original input material to the digesters that the microbes cannot use. It also consists of the mineralised remains of the dead bacteria from within the digesters. Digestate can come in three forms: fibrous, liquor, or a sludge-based combination of the two fractions. In two-stage systems, different forms of digestate come from different digestion tanks. In single-stage digestion systems, the two fractions will be combined and, if desired, separated by further processing.

Acidogenic anaerobic digestate

The second byproduct (acidogenic digestate) is a stable, organic material consisting largely of lignin and cellulose, but also of a variety of mineral components in a matrix of dead bacterial cells; some plastic may be present. The material resembles domestic compost and can be used as such or to make low-grade building products, such as fibreboard. The solid digestate can also be used as feedstock for ethanol production.

The third byproduct is a liquid (methanogenic digestate) rich in nutrients, which can be used as a fertiliser, depending on the quality of the material being digested. Levels of potentially toxic elements (PTEs) should be chemically assessed. This will depend upon the quality of the original feedstock. In the case of most clean and source-separated biodegradable waste streams, the levels of PTEs will be low. In the case of wastes originating from industry, the levels of PTEs may be higher and will need to be taken into consideration when determining a suitable end use for the material.

Digestate typically contains elements, such as lignin, that cannot be broken down by the anaerobic microorganisms. Also, the digestate may contain ammonia that is phytotoxic, and may hamper the growth of plants if it is used as a soil-improving material. For these two reasons, a maturation or composting stage may be employed after digestion. Lignin and other materials are available for degradation by aerobic microorganisms, such as fungi, helping reduce the overall volume of the material for transport. During this maturation, the ammonia will be oxidized into nitrates, improving the fertility of the material and making it more suitable as a soil improver. Large composting stages are typically used by dry anaerobic digestion technologies.

Wastewater

The final output from anaerobic digestion systems is water, which originates both from the moisture content of the original waste that was treated and water produced during the microbial reactions in the digestion systems. This water may be released from the dewatering of the digestate or may be implicitly separate from the digestate.

The wastewater exiting the anaerobic digestion facility will typically have elevated levels of biochemical oxygen demand (BOD) and chemical oxygen demand (COD). These measures of the reactivity of the effluent indicate an ability to pollute. Some of this material is termed 'hard COD', meaning it cannot be accessed by the anaerobic bacteria for conversion into biogas. If this effluent were put directly into watercourses, it would negatively affect them by causing eutrophication. As such, further treatment of the wastewater is often required. This treatment will typically be an oxidation stage wherein air is passed through the water in a sequencing batch reactors or reverse osmosis unit.

History

Gas street lamp

Reported scientific interest in the manufacturing of gas produced by the natural decomposition of organic matter dates from the 17th century, when Robert Boyle (1627-1691) and Stephen Hales (1677-1761) noted that disturbing the sediment of streams and lakes released flammable gas. In 1808 Sir Humphry Davy proved the presence of methane in the gases produced by cattlemanure. In 1859 a leper colony in Bombay in India built the first anaerobic digester. In 1895, the technology was developed in Exeter, England, where a septic tank was used to generate gas for the sewer gas destructor lamp, a type of gas lighting. Also in England, in 1904, the first dual-purpose tank for both sedimentation and sludge treatment was installed in Hampton, London. In 1907, in Germany, a patent was issued for the Imhoff tank, an early form of digester.

Research on anaerobic digestion began in earnest in the 1930s.

Incineration: An Introduction

Incineration is a chemical reaction in which carbon, hydrogen and other elements in the waste mix with oxygen in the combustion zone and generates heat. The air requirements for combustion of solid wastes are considerable. For example, approximately 5000 kg of air is required for each tonne of solid wastes burned. Usually, excess air is supplied to the incinerator to ensure complete mixing and combustion and to regulate operating temperature and control emissions. Excess air requirements, however, differ with moisture content of waste, heating values and the type of combustion technology employed. The principal gas products of combustion are carbon dioxide, carbon monoxide, water, oxygen and oxides of nitrogen.

Many incinerators are designed to operate in the combustion zone of 900°C – 1100°C. This temperature is selected to ensure good combustion, complete elimination of odours and protection of the walls of the incinerator. Incinerator systems are designed to maximise waste burn out and heat output, while minimising emissions by balancing the oxygen (air) and the three "Ts", i.e., time, temperature and turbulence. Complete incineration of solid wastes produces virtually an inert residue, which constitutes about 10% of the initial weight and perhaps a larger reduction in volume. The residue is generally landfilled.

The incineration facility along with combustion of waste emits air pollutants (i.e., fine particulate and toxic gases), which are an environmental concern, and, therefore, their control is necessary. Other concerns relating to incineration include the disposal of the liquid wastes from floor drainage, quench water, scrubber effluents and the problem of ash disposal in landfills because of heavy metal residues. By optimising the combustion process, we can control the emission of combustible, carbon-containing pollutants (EPA 1989 and 1995). Oxides of nitrogen and sulphur, and other gaseous pollutants are not considered a problem because of their relatively smaller concentration.

Combustion of Waste Material

Table shows the major elements that constitute solid wastes and the end products of combustion:

Major Elements of Solid Wastes

Elements		End Products (Ga se s)
Carbon (C)		Carbon dioxide (CO_2)
Hydrogen (H)		Water (H_2O)
Oxygen (O)	Combustion Process	Nitrogen (N2)
Nitrogen (N)		
Sulphur (S)		Sulphur dioxide (SO_2), other gaseous compounds and ash

Table gives the information about several components of solid waste mixtures on the basis of proportion:

Ultimate Analysis of Combustible Component

Component	Percent by Weight (dry basis)					
	Carbon	Hydrogen	Oxygen	Nitrogen	Sulphur	Ash
Food waste	48.0	6.4	37.6	2.6	0.4	5.0
Paper	43.5	6.0	44.0	0.3	0.2	6.0
Cardboard	44.0	5.9	44.6	0.3	0.2	6.0
Plastic	60.0	7.2	22.8	--	--	10.0
Textile	55.0	6.6	31.2	4.6	0.15	2.5
Rubber	78.0	10.0	--	2.0	--	10.0
Leather	60.0	8.0	11.6	10.0	0.4	10.0
Garden trimmings	47.8	6.0	38.0	3.4	0.3	4.5
Wood	49.5	6.0	42.7	0.2	0.1	1.
Dirt, ash, brick, etc.	26.3	3.0	2.0	0.5	0.2	68.0

In case energy values in KJ/kg or BTU/1b are not available, we can calculate them approximately from the data in Table above and the Dulong formula given below:

$$\text{Energy value (BTU/lb)} = 145.4\,C + 620\,(H - 1/8\,O) + 41S$$

where C, H, O, and S are in percent by weight (dry basis) and can be converted to KJ/kg by: BTU/lb x 2.326 = KJ/kg

Incineration Objectives

The purpose of incineration is to combust solid wastes to reduce their volume to about one-tenth, without producing offensive gases and ashes (Phelps, et al., 1995). That is to say, incineration of solid wastes aims at the following (McDougall, et al., 2001):

- Volume reduction: Depending on its composition, incineration reduces the volume of solid wastes to be disposed of by an average of 90%. The weight of the solid wastes to be dealt with is reduced by 70 – 75%. This has both environmental and economic advantages since there is less demand for final disposal to landfill, as well as reduced costs and environmental burdens due to transport, if a distant landfill is used.

- Stabilisation of waste:Incinerator output (i.e., ash) is considerably more inert than incinerator input (i.e., solid wastes), mainly due to the oxidation of the organic components of the waste stream. This leads to a reduction of landfill management problems (since the organic fraction is responsible for landfill gas production) and the organic compounds present in landfill leachate.

- Recovery of energy from waste (EFW): This represents a valorisation method, rather than just a pre-treatment of waste prior to disposal. Energy recovered from burning the wastes is used to generate steam for use in on- site electricity generation or export to local factories or district heating schemes. Combined heat and power plants increase the efficiency of energy recovery by producing electricity as well as utilising the residual heat. Solid waste incineration can replace The prevalence of incineration practice and the actual approach that it takes vary across regions and reflect the relative importance attached to the different objectives discussed. Volume reduction for both environmental and economic reasons and sterilisation of waste has historically been the important objectives of incineration. Due to the growing concern over the production of landfill gas and the organic compounds in leachate from landfill receiving untreated waste, it is likely that the future will see more emphasis on using incineration for stabilising wastes for subsequent landfilling. Landfill gas and leachate arise principally from the organic fraction of solid wastes, which will be effectively converted to gases and mineralised ash by incineration.the use of fossil fuels for energy generation. As a large part of the energy content of solid wastes comes from truly renewable resources (e.g., biomass), there should be a lower overall net carbon dioxide production than that from burning fossil fuels, since carbon dioxide is absorbed in the initial growing phase of the biomass.

- Sterilisation of waste: This is of primary importance in the incineration of clinical or biomedical waste. Incineration of solid wastes will also ensure destruction of pathogens prior to final disposal in a landfill.

Planning an Incineration Facility

Incineration of solid wastes is becoming an increasingly important aspect of solid waste management, as communities look for alternatives to rapidly filling landfills (or disappearing landfill sites). Modern incineration facilities are no longer simple garbage burners. Instead, they are designed to produce steam and electricity and can be used as a complement to source reduction, recycling and composting programmes. However, strategic long- term planning is essential for developing a successful incineration facility. In other words, it is important to develop an understanding of a variety of issues in the planning process, including the following:

(i) Facility ownership and operation: One of the first planning decisions that local officials make is about the entity that will actually own the facility and oversee its operation. This decision is based largely on the amount of financial risk the community is willing to assume and the time and resources available. Some of the procurement options, in this context, are:

- Full service approach: In this system, the community specifies only the process type and the performance required, and hires a (single) firm to design, construct and operate the plant.

- Merchant plants: In this type of system, in which waste is accepted on weight basis, a private firm designs, constructs, owns and operates the facility.

- Turnkey approach: In this system, a single company designs and builds the plant, according to the communities' specifications, and the community or a different contractor owns and operates the plant.

(ii) Energy market: Waste incineration facilities differ from most government services in that they generate a product as well as energy, which are sold for revenue. Steam and electricity are the energy products at incineration facilities, depending on the particular design.

(iii) Marketing steam: The primary end uses of steam from waste incineration facilities are industrial and institutional heating and cooling systems. Marketing of steam products involves identifying these industries and institutions within the region. Industrial and institutional steam users include textile, paper and pulp, food processing, leather, chemical producers, hospitals, etc. Planning must include proper backup to guarantee a consistent supply and steam demand variation (often caused by changing seasons).

(iv) Marketing electricity: Incineration facilities generating electricity are referred to as co-generators as they provide electricity in addition to that generated by the local electric utility. Besides the plant that uses the electricity generated for its operation, customers for electricity include nearby industries and public, and private utilities. It must be equipped to give a consistent supply and mustcompete with other co-generators in selling energy.

(v) Facility siting: Siting the incineration facility is one of the most important tasks to be undertaken and a variety of social and technical hurdles have to be negotiated. The important aspects in this context are the following:

- Effect on residents: Residents will be concerned with the health effects associated with incinerator plant, decreased property value and increased traffic (e.g., due to truck movement).

- Environmental impact: Incineration has the potential tocreatea variety of environmental concerns like air, water and noise pollution and ash disposal.

- Development plans: It is important to evaluate future land use plans at the possible site.

- Proximity towastesource: Transportation cost isone of the most significant expenditures in waste management system.

- Proximity toenergy market: The energy products will have to be delivered to buyers. The location of power line must be considered.

- Logistic concerns: Area zoning and access route must be considered.

- Residual ash disposal: Access to a secure landfill is necessary.

(vi) Facility sizing: Proper plant sizing results from careful evaluation of a wide variety of criteria such as:

- Waste supply: This is the most fundamental sizing factor and measures are usually taken to guarantee a waste supply for the facility. Waste flow control ordinances are often used to ascertain the quantity of waste. When properly planned, waste flow control can benefit both incineration facility and alternative waste management programme, by diverting the relevant portions of the waste stream (e.g., recyclables to the recycling programme and combustibles to incineration facility).

- Alternative waste management programme: In addition to waste flow control agreements, future source reduction, recycling and composting programmes are directly related to facility design. When sizing the incineration facility, it is important, therefore, to account for the type and amount of materials that will be diverted from the facility.

- Waste stream characteristics: Good combustion depends on the accuracy of waste stream data. Planning of incineration facility requires waste stream assessment to develop an accurate picture of the quantity and composition of the waste stream. From a technical standpoint, the waste streamdata will be used to ascertain the heating value of the waste, which helps in plant operation.

- Planning for facility disruption: Accounting for downtime is an important facility planning criteria. Most incinerationfacilitiesare designed to operate continuously (i.e., 24 hours a day), but both scheduled (e.g., maintenance)andunscheduled(e.g.,equipment failure) downtime situations are likely to occur. Storage space must be available for the waste that continues to arrive during downtime. If these capabilities are not built into the system, provisions must be made to send waste to a landfill or an alternative facility.

- Facility financing: Depending upon the procurement approach selected, incineration facility will require extensive financing agreements.

- Time frame: The time required to plan, develop and construct a facility will vary, but at least 5 to 8 years are required to bring a new facility from the early planning stages to in-service.

Long-term planning within the local government is the key for successful facility design and operation. By understanding all issues anddedicatedworkforce, waste combustion can become a positive component of waste management system (EPA 1989 and 1995).

Incineration Technologies

The four incineration technologies covered in this section are mass burning system, refuse derived fuel system, modular incineration and fluidised bed incineration. The two most widely used and technically proven incineration technologies are mass-burning incineration and modular incineration. Fluidised- bed incineration has been employed to a lesser extent, although its use has been expanding and experience with this relatively new technology has increased. Refuse-derived fuel production and incineration has also been used, with limited success. Some facilities have been used inconjunctionwithpyrolysis, gasification and other related processes that convert solid waste to gaseous, liquid, or solid fuel through thermal processing (UNEP 1996).

Mass-burning System

Mass-burning systems are the predominant form of MSW incineration. A mass- burn facility typically consists of a reciprocating grate combustion system and a refractory-lined, water-walled steam generator. Mass-burn systems generally consist of either two or three incineration units ranging in capacity from50 to 1,000 tonnes per day. That is to say, the facility capacity ranges from about 100– 150 to 2,000 – 3,000 tonnes per day. These facilities can accept refuse that has undergone little preprocessing other than the removal of oversized items. Although this versatility makes mass-burn facilities convenient and flexible, local programmes to separate household hazardous wastes (e.g., cleaners and pesticides) and recover certain materials (e.g., iron scrap) are necessary to help ensure environmentally viable incineration and resource conservation.

Because of the larger facility size, an incineration unit is specially designed to efficiently combust the waste to recover greater quantities of steam or electricity for revenue. To achieve this greater combustion and heat recovery efficiency, the larger field-erected incinerators are usually in-line furnaces with a grate system. The steam generator generally consists of refractory-coated water wall systems,i.e., walls comprised of tubes through which water circulates to absorb the heat of combustion. In a water wall system, the boiler is an integral part of the system wall, rather than a separate unit as is in a refractory system.

Mass-burning of waste can also be achieved by the use of a rotary kiln. Rotary kilns use a turning cylinder, either refractor or water wall design, to tumble the waste through the system. The kiln is reclined, with waste entering at the high elevation end and ash and non-combustibles leaving at the lower end.

The waste intake area usually includes a tipping floor, pit, crane and sometimes conveyors. Trucks enter the tipping floor and tip their wastes either onto the floor itself, or directly into the pit. When wastes are tipped onto the floor, a front-end loader or a bulldozer is used to push them into the pit or onto a conveyor. From a feed chute, MSW is continuously fed to a grate system, which moves the waste through a combustion chamber using a tumbling motion. A travelling or reciprocating grate may follow rotary combustors to further complete combustion.

At least two combustor units are included to provide a level of redundancy and to allow waste processing at a reduced rate during periods of scheduled and unscheduled maintenance. Mass-burn facilities today generate a higher quality of steam (i.e., pressure and temperature), which is then passed through a turbine generator to produce electricity or through an extraction turbine to generate electricity as well as provide process steam for heating or other purposes.

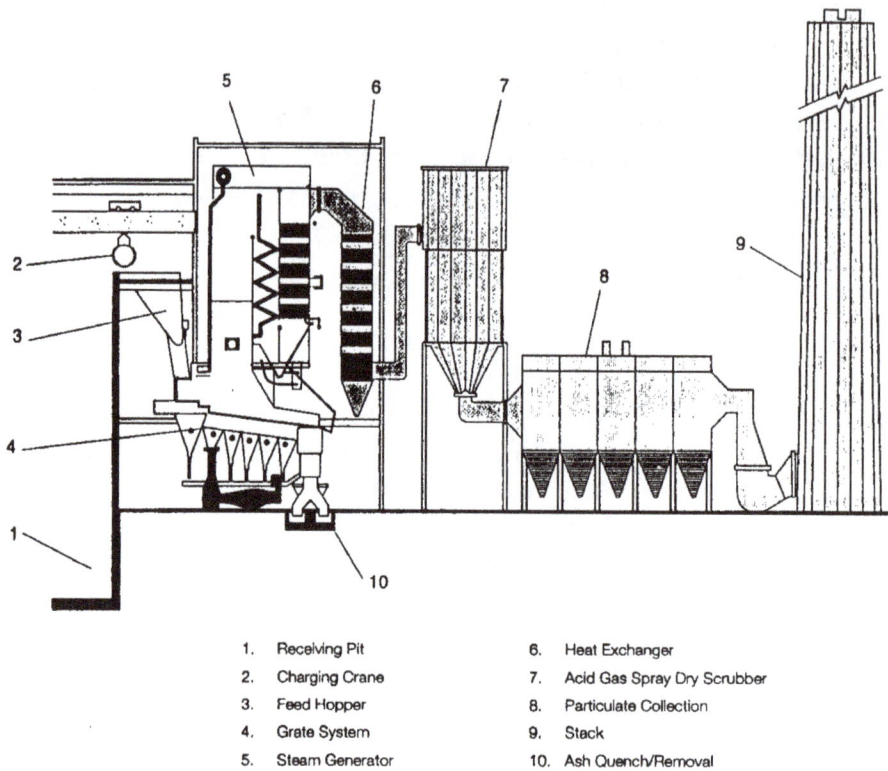

1.	Receiving Pit	6.	Heat Exchanger
2.	Charging Crane	7.	Acid Gas Spray Dry Scrubber
3.	Feed Hopper	8.	Particulate Collection
4.	Grate System	9.	Stack
5.	Steam Generator	10.	Ash Quench/Removal

Typical Mass-Burn Facility

Refuse Derived Fuel (RDF) System

Refuse-derived fuel (RDF) refers to solid wastes in any form that is used as fuel. The term RDF, however, is commonly used to refer to solid waste that has been mechanically processed to produce a storable, transportable and more homogeneous fuel for combustion. RDF systems have two basic components: RDF production and RDF incineration.

RDF production facilities make RDF in various forms through material separation, size reduction and pelletising. Although RDF processing has the advantage of removing recyclables and contaminants from the combustion stream, on an average, capital costs per tonne for incineration units that use RDF are higherthan for other incineration options. RDF production plants like mass-burn incinerators characteristically have an indoor tipping floor. Instead of being pushed onto a pit, the waste in an RDF plant is typically fed onto a conveyor, which is either below grade or hopper fed. In some plants, the loader doing the feeding will separate corrugated and bulky items, like carpets.

Once on the conveyor, the waste travels through a number of processing stages, usually beginning with magnetic separation. The processing steps are tailored to the desired products, and typically include one or more screening stages, using trammel or vibrating screens, shredding or hammer milling of waste with additional screening steps, pelletising or baling of burnable wastes, and, depending on the local recycling markets and the design of the facility, a manual separation line.

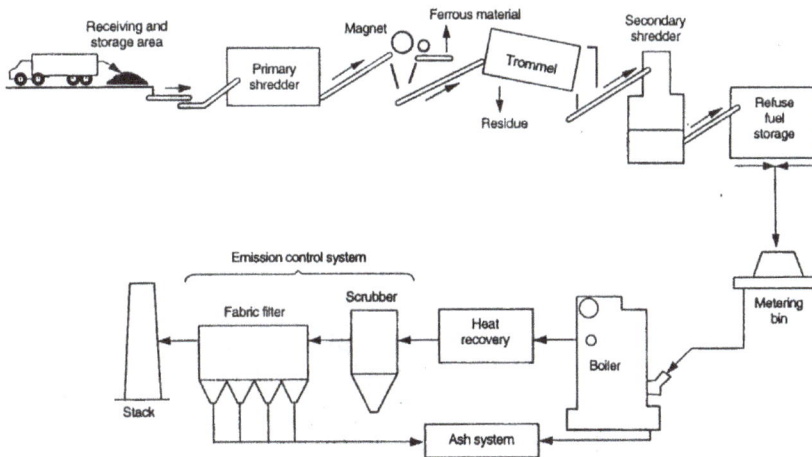

Typical Simplified RDF Facility

There are two primary types of systems in operation, and these are:

(i) Shred-and-burn systems: Shred-and-burn systems are the simplest form of RDF production. The process system typically consists of shredding the MSW to the desired particle size that allows effective feeding to the combustor and magnetic removal of ferrous metal, with the remaining portion delivered to the combustor. There is no attempt to remove other non-combustible materials in the MSWbeforecombustion.This,in essence, is a system with minimal processing and removal of non- combustibles.

(ii) Simplified process systems: This is a system that removes a significant portion of the non-combustibles. A simplified process system involves processing the MSW to produce an RDF with a significant portion of the non-combustibles removed before combustion. The MSW process removes more than 85% of the ferrous metals, a significant percentage of the remaining non-combustible (i.e., glass, nonferrous metals, dirt, sand, etc.), and shreds the material to a nominal particle top size of 10 to 15 cm to allow effective firing in the combustion unit.

Depending on the type of combustor to be used, a significant degree of separation can be achieved to produce a high-quality RDF (i.e., low ash), which typically results in the loss of a higher percentage of combustibles when compared to systems that can produce a low-quality fuel (i.e., slightly higher ash content) for firing in a specially designed combustor. These types of systems recover over 95% of the combustibles in the fuel fraction.

Modular Incineration

Modular incinerator units are usually prefabricated units with relatively small capacities between 5 and 120 tonnes of solid waste per day. Typical facilities have between 1 and 4 units with a total plant capacity of about 15 to 400 tonnes per day. The majority of modular units produce steam as the soleenergy product. Due to their small capacity, modular incinerators are generally used in small communities or for commercial and industrial operations.

Their prefabricated design gives modular facilities the advantage of a shorter construction time. Modular combustion systems are usually factory-assembled units consisting of a refractory-lined furnace and a waste heat boiler. Both units can be pre-assembled and shipped to the construction site, which minimises field installation time and cost. Adding modules or units, installed in parallel can increase facility capacity. For example, a 200 tonne-per-day facility may consist of 4 units, a 50-tonne-per-day consists of 2 units and a 100tonne-per-day consists of 1 unit. The number of units may depend on the fluctuation of waste generation for the service area and the anticipated maintenance cycle of the units.

Modular incinerators employ a different process from that of mass-burn incinerators, typically involving two combustion chambers, and combustion is typically achieved in two stages.

The first stage may be operated in a condition in which there is less than the theoretical amount of air necessary for complete combustion. The controlled aircondition creates volatile gases, which are fed intothesecondarychamber, mixed with additional combustion air, and under controlled conditions, completely burned. Combustion temperatures in the secondary chamber are regulated by controlling the air supply, and when necessary, through the use of an auxiliary fuel. The hot combustion gases then pass through a waste heat boiler to produce steam for electrical generation or for heating purposes. The combustion gases and products are processed through air emission control equipment to meet the required emission standards.

In general, modular incineration systems are a suitable alternative and may, for smaller-sized facilities, be more cost-effective than other incinerators. But modular incineration has become less common, partly due to concerns over the consistency and adequacy of air pollution controls.

Fluidised-bed Incineration

Fluidised-bed incineration of MSW is typically medium scale, with processing capacity from 50 to 150 tonnes per day. In this system, a bed of limestone or sand that can withstand high temperatures, fed by an air distribution system, replaces the grate. The heating of the bed and an increase in the air velocities cause the bed to bubble, which gives rise to the term fluidised. There are two types of fluidised-bed technologies, viz., bubbling bed and circulating bed. The differences are reflected in the relationship between air flow and bed material, and have implications for the type of wastes that can be burned, as well as the heat transfer to the energy recovery system.

Unlike mass-burn incinerators, fluidised-bed incinerators require front-end pre- processing, also called fuel preparation. They are generally associated with source separation because glass and metals do not fare well in these systems and also they can successfully burn wastes of widely varying moisture and heat content, so that the inclusion of paper and wood, which are both recyclable and burnable, is not a crucial factor in their operation (and thus paper can be extracted for higher-value recycling).

Fluidised-bed systems are more consistent in their operation than mass burn and can be controlled more effectively to achieve higher energy conversion efficiency, less residual ash and lower air emissions. In general, however, these systems appear to operate efficiently on smaller scales thanmass-burnincinerators, which may make them attractive in some situations. For this reason, fluidised- bed technology may be a sound choice for high-recycling cities in developing countries when they first adopt incineration.

Indian Scenario in Selection of Incineration Technology

The absence of a well planned, scientific system of waste management (including waste segregation at source) coupled with ineffective regulation leading to waste burning. The left-over waste at the dumping yards generally contains high percentage of inerts (>40%) and of puterscible organic matter (30- 60%). It is common practice of adding the road sweepings to the dust bins. Papers and plastics are mostly picked up and only such fraction which is in an unrecoverable form remains in the refuse. Paper normally constitutes 3-7% of refuse while the plastic content is normally less than 1%. The calorific value on dry weight basis (High Calorific Value) varies between 800-1100 k-cal/kg. Self sustaining combustion cannot be obtained for such waste and auxiliary fuel will be required. Incineration, therefore, has not been preferred in India so far. The only incineration plant installed in the country is at Timarpur, Delhi way back in the year 1990 has been lying inoperative due to mismatch between the available waste quality and plant design. This made the government of Delhi to assure increased efforts in segregation of household MSW at source collection. However, with the growing problems of waste management in the urban areas and the increasing awareness about the ill effects of the existing waste management practices on the public heath, the urgent need for improving the overall waste management system and adoption of advanced, scientific methods of waste disposal, including incineration, is imperative.

Out of most recent Waste to energy technologies adopted in India such as Biomethanation, landfill with gas recovery, gasification/pyrolysis, incineration and composting; incineration is selected as lastoption.

The benefit of incineration is a substantial reduction in the weight and volume of waste, and generation of revenue from energy production known as "waste-to- energy" (WTE), which can partially offset the cost of incineration. Keeping this in view, we will discuss the various options of energy generation from waste.

References

- Aslam DN, et al. "Development of models for predicting carbon mineralization and associated phytotoxicity in compost-amended soil.". Bioresour Technol. 99: 8735–41. PMID 18585031. doi:10.1016/j.biortech.2008.04.074

- Haughey, A. (1968). "The Planktonic Algae of Auckland Sewage Treatment Ponds". New Zealand Journal of Marine and Freshwater Research. 2 (4): 721–766. doi:10.1080/00288330.1968.9515271

- Martin V. Melosi (2010). The Sanitary City: Environmental Services in Urban America from Colonial Times to the Present. University of Pittsburgh Press. p. 110. ISBN 9780822973379

- "Lowering Cost and Waste in Flue Gas Desulfurization Wastewater Treatment". Power Mag. Electric Power. Retrieved 6 April 2017

- Ashton, John; Ubido, Janet (1991). "The Healthy City and the Ecological Idea" (PDF). Journal of the Society for the Social History of Medicine. 4 (1): 173–181. doi:10.1093/shm/4.1.173. Retrieved 8 July 2013

- Colin A. Russell (2003). Edward Frankland: Chemistry, Controversy and Conspiracy in Victorian England. Cambridge University Press. pp. 372–380. ISBN 9780521545815

- "Waste Pesticide Management" (PDF). Oregon.Gov. State of Oregon Department of Environmental Quality Land Quality Division Hazardous Waste Program. Retrieved 5 October 2016

- Sharma, Sanjay Kumar; Sanghi, Rashmi (2012). Advances in Water Treatment and Pollution Prevention. Springer. ISBN 9789400742048. Retrieved 2013-02-07

- Courtney Symons (13 October 2011). "'Humanure' dumping sickens homeowner". YourOttawaRegion. Metroland Media Group Ltd. Retrieved 16 October 2011

- Roubík, Hynek; Mazancová, Jana; Banout, Jan; Verner, Vladimír (2016-01-20). "Addressing problems at small-scale biogas plants: a case study from central Vietnam". Journal of Cleaner Production. 112, Part 4: 2784–2792. doi:10.1016/j.jclepro.2015.09.114

- Tilley, David F. (2011). Aerobic Wastewater Treatment Processes: History and Development. IWA Publishing. ISBN 9781843395423. Retrieved 2013-02-07

- Lindsay, Jay (12 June 2012). "Japanese composting may be new food waste solution". AP. Retrieved 13 November 2012

- Benner, Ronald (1989). "Book Review: Biology of anaerobic microorganisms" (PDF). Limnology and Oceanography. 34 (3): 647. doi:10.4319/lo.1989.34.3.0647. Archived from the original (PDF) on 13 November 2006

- Tchobanoglous, G., Burton, F.L., and Stensel, H.D. (2003). Wastewater Engineering (Treatment Disposal Reuse) / Metcalf & Eddy, Inc. (4th ed.). McGraw-Hill Book Company. ISBN 0-07-041878-0

- "Electronic Code of Federal Regulations. Title 40, part 503. Standards for the use or disposal of sewage sludge". U.S. Government Printing Office. 1998. Retrieved 30 March 2009

- Jewell, W.; Cummings, R.; Richards, B. (1993). "Methane fermentation of energy crops: Maximum conversion kinetics and in situ biogas purification". Biomass and Bioenergy. 5 (3–4): 261–278. doi:10.1016/0961-9534(93)90076-G

- Hemenway, Toby (2009). Gaia's Garden: A Guide to Home-Scale Permaculture. Chelsea Green Publishing. pp. 84-85. ISBN 978-1-60358-029-8

- National Non-Food Crops Centre. "NNFCC Renewable Fuels and Energy Factsheet: Anaerobic Digestion", Retrieved on 2011-11-22

- Feasibility study concerning anaerobic digestion in Northern Ireland, eunomia.co.uk , Retrieved 19.08.07. Archived 28 November 2007 at the Wayback Machine

- Zehnder, Alexander J. B. (1978). "Ecology of methane formation". In Mitchell, Ralph. Water pollution microbiology 2. New York: Wiley. pp. 349–376. ISBN 978-0-471-01902-2

- Pamatmat, Mario Macalalag; Bhagwat, Ashok M. (1973). "Anaerobic metabolism in Lake Washington sediments" (PDF). Limnology and Oceanography. pp. 611–627. doi:10.4319/lo.1973.18.4.0611. Archived (PDF) from the original on 16 December 2013

Permissions

All chapters in this book are published with permission under the Creative Commons Attribution Share Alike License or equivalent. Every chapter published in this book has been scrutinized by our experts. Their significance has been extensively debated. The topics covered herein carry significant information for a comprehensive understanding. They may even be implemented as practical applications or may be referred to as a beginning point for further studies.

We would like to thank the editorial team for lending their expertise to make the book truly unique. They have played a crucial role in the development of this book. Without their invaluable contributions this book wouldn't have been possible. They have made vital efforts to compile up to date information on the varied aspects of this subject to make this book a valuable addition to the collection of many professionals and students.

This book was conceptualized with the vision of imparting up-to-date and integrated information in this field. To ensure the same, a matchless editorial board was set up. Every individual on the board went through rigorous rounds of assessment to prove their worth. After which they invested a large part of their time researching and compiling the most relevant data for our readers.

The editorial board has been involved in producing this book since its inception. They have spent rigorous hours researching and exploring the diverse topics which have resulted in the successful publishing of this book. They have passed on their knowledge of decades through this book. To expedite this challenging task, the publisher supported the team at every step. A small team of assistant editors was also appointed to further simplify the editing procedure and attain best results for the readers.

Apart from the editorial board, the designing team has also invested a significant amount of their time in understanding the subject and creating the most relevant covers. They scrutinized every image to scout for the most suitable representation of the subject and create an appropriate cover for the book.

The publishing team has been an ardent support to the editorial, designing and production team. Their endless efforts to recruit the best for this project, has resulted in the accomplishment of this book. They are a veteran in the field of academics and their pool of knowledge is as vast as their experience in printing. Their expertise and guidance has proved useful at every step. Their uncompromising quality standards have made this book an exceptional effort. Their encouragement from time to time has been an inspiration for everyone.

The publisher and the editorial board hope that this book will prove to be a valuable piece of knowledge for students, practitioners and scholars across the globe.

Index